Recent Advances in Green Technologies and Sustainable Development

Edited By

Dr. Mahesh M Bundele, Dr. Rekha Nair and Dr. Piyusha Somvanshi

Book description

Recent advances in green technologies and sustainable development deals with cutting-edge research and innovative ideas in different categories of green technologies and operational aspects of sustainable development including renewable energy sources, power systems, mathematical ecology, industrial technologies, construction and material sciences.

The chapters are written by eminent and insightful authors to propose improvement and expansion of processes and applications connected to sustainable development. Environmental awareness and protection are one of the challenging issues of the new millennia. Industrialization and population explosion has opened new frontiers in the conservation of environmental protection. Rapid urbanization is proving to have direct consequences on the environment. The need of the hour is a balanced approach to multi layered conservative methods. Any sustainable development has a multifaceted approach, encompassing environmental, technological, social, and economical developmental dimensions. This book focuses on these various issues in a progressive manner. The selected papers in this book have highlighted a plethora of issues related to green technology and sustainable development. Ample care has been given to selecting the papers which tried to bridge the gap between technological advancement and its impact on the environment.

Editor(s)
Biography

Dr. Mahesh M Bundele

Dr. Mahesh M Bundele is working as Principal and Director of Poornima College of Engineering, Jaipur, India. He completed his doctorate from Amaravati University in Computer Science & Engineering with a topic "Design and Implementation of Wearable Computing System for the

Prevention of Road Accidents". While working he has handled various portfolios related to design and development of curriculum, design and execution of laboratories, electrical installations including substation design and erection, street lights etc. He has worked on various Wi-Fi/Wi-Max based system designs and implementations for college and Pusad city villages. He has guided many research projects at UG and PG level on various applications in Electrical, Electronics & Computer Sciences. He has designed and executed various LT & HT electrical installations. He has visited the US, UK, China and Malaysia for research presentations. He has delivered many invited talks and has been on advisory committees of IEEE International Conferences. He has published more than 70 research papers in National and International conferences and journals. He is a senior member of IEEE & ACM, life member of ISTE and IEI. He has worked as Secretary in IEEE Rajasthan Subsection and also as Membership Development Chair in IEEE Rajasthan Subsection. He is a member standing committee member for Technical and Professional activities at IEEE Delhi Section and also a member of EXECOM IEEE Delhi section. Currently he is Chair in Jaipur ACM Professional Chapter and plays an active role in ACM activities. He is working on research projects on Indian Rail Crack Detection and Monitoring and Micro-grid design and implementation in un-electrified villages of Rajasthan. His areas of research interest are Wearable & Pervasive Computing, Artificial Intelligence, Software Defined Networking, Wireless Sensor Networks, Smart Grids, Micro Grids etc. He has organized and chaired many IEEE, ACM, Elsevier and Springer international conferences and workshops and delivered keynote addresses.

Dr. Rekha Nair

Dr. Rekha Nair is a Professor and Dean Academics at Poornima Group. She is PhD in "Corrosion Inhibition for Aluminium and its Alloys by Some Natural Products" from University of Rajasthan, Jaipur. Her research areas are corrosion inhibition for metals, surface coatings, electrochemistry and environmental pollution. She has over two decades of experience in teaching and administration with more than fifty publications in national/international journals and conferences. As a renowned book author of six books, her books are referred to in several universities as textbooks. She has been the author of many book chapters of international repute. While working, she has handled various portfolios related to design and development of curriculum, design and execution of laboratories. She is life member of ISTE and Electrochemical Society of India, IISc Bengaluru. She has organized a number of workshops, conferences and training at national and international level. Her recent contribution as an editor and reviewer was for Scopus Indexed IOP Conference Series: Earth and Environmental Science (795) - International Conference on Sustainable Energy, Environment and Green Technologies, 2021. She has been a reviewer for international journals such as Corrosion Science, Pigment and Resin Technology and several international conferences. She has been an invited speaker and chaired the sessions at National/ International conferences.

Dr. Piyusha Somvanshi

Dr. Piyusha Somvanshi is presently working as a Professor in the Department of Applied Sciences at Poornima College of Engineering, Jaipur, India. She is MSc. (Gold Medalist) in Applied Mathematics and pursued her PhD from National Institute of Technology, Raipur, India. She has 15 years of teaching experience at University level. She has 20+ publications in SCI and SCOPUS indexed journals and in national and international conferences. She is member of Journal of Korean Mathematical Society ISSN 1225-1763 (print), ISSN 2234-3024 (online) and she has been a reviewer for Scopus Indexed Journal Advances in Operator's Theory, Tusi Mathematical Research Group (TMRG), (e-ISSN 2538-225X), Iran. She is also an Udemy Instructor and member of Academic Council in Apex University, Jaipur, India.

Recent Advances in Green Technologies and Sustainable Development

[First edition]

Edited By

Dr. Mahesh M Bundele

Dr. Rekha Nair
ORCID 0000-0002-6983-0518

Dr. Piyusha Somvanshi
ORCID 0000-0001-5381-7610

CRC Press
Taylor & Francis Group
Boca Raton London New York

CRC Press is an imprint of the
Taylor & Francis Group, an **informa** business

First edition published 2024
by CRC Press
4 Park Square, Milton Park, Abingdon, Oxon, OX14 4RN

and by CRC Press
2385 NW Executive Center Drive, Suite 320, Boca Raton FL 33431

CRC Press is an imprint of Informa UK Limited

British Library Cataloguing-in-Publication Data
A catalogue record for this book is available from the British Library

ISBN: 9781032586465 (pbk)
ISBN: 9781003450917 (ebk)

DOI: 10.1201/9781003450917

Typeset in Sabon LT Std
by HBK Digital

Printed and bound in India

Contents

List of Figures

List of Tables

Acknowledgement

The world is a better place to thank people who want to develop and lead others. What makes it even better are people who share the gift of their time and knowledge to mentor others. Having an idea and turning it into a book is both internally challenging and rewarding.

We extend our sincere thanks to Ar. Rahul Singhi, Director, Poornima Group for his continuous support. We are grateful to Vice Principal Dr. Pankaj Dhemla, Head of the departments at Poornima College of Engineering Dr. Gajendra Singh, Dr. Praveen Sonwane, Dr. Pran Dadhich, Dr. Garima Mathur, Dr. Nikita Jain and the working team without their experience and support it would not be possible for us. We thank our team and peers for their efforts and technical support.

We offer our sincere thanks to whole Taylor and Francis Group, Dr. Gagandeep Singh, Senior Publisher Taylor and Francis group, Mr. Praveen Singh Dewal, Senior Territory Manager, Taylor and Francis group and Ms. Afreen Ayub for their constant support and always being there to resolve our queries.

We would like to acknowledge our family with gratitude for their love and support. They kept us going, without their support it would not have been possible to complete the book.

Last but not the least we are grateful to all of those who have provided us with their extensive personal and professional guidance and support.

1 Analysing industry 4.0 key performance indicators for sustainability of smart warehouses

Vijay Prakash Sharma[1,a], Surya Prakash[2],
Ranbir Singh[1] and Chirag Malik[3]

[1]School of Engineering and Technology, BML Munjal University, Gurugram, Haryana (India)

[2]Operations Management Department, Great Lakes Institute of Management, Gurgaon, Haryana (India)

[3]School of Management, BML Munjal University, Gurgaon, Haryana (India)

Abstract

A warehouse is essential to the supply chain network and product delivery system. Economic development, humanitarian aspects, social consequences, and environmental concerns are the four fundamental pillars of sustainability. Industry 4.0 (I4.0) technologies influence all four measuring pillars of sustainable development. The study focuses on measuring the impact of I4.0 key performance indicators (KPIs) on sustainable warehouse logistics networks. The literature review and expert discussion are carried out to identify and establish the KPIs for smart warehousing. The intelligent supply chain network's KPIs are formulated using analytical hierarchy process (AHP). The adoption of IoT-based RFID systems and AI-based machine learning technologies are the prominent KPIs for achieving the goal of sustainability in the smart warehouse. There is a need to have real-time data for establishing a connection between I4.0 and warehouse sustainability. The specific firm-based case studies can help to establish frameworks for desired logistic networks. KPIs for efficient WMS are prioritized for achieving goals of sustainability. The research provides a guideline by explaining the concept of I4.0 for smart warehousing.

Keywords: Artificial intelligence, industry 4.0, Internet of Things, sustainability, warehouse management system (WMS).

Introduction

Industry 4.0 (I4.0) is a technological revolution that uses smart technologies to automate, monitor, and analyze supply chain networks. Internet of Things (IoT), artificial intelligence (AI), and cyber-physical systems, are intelligent, integrated systems that control and monitor physical items. It helps to operate machinery, robots, drones, automated vehicles, and equipment using machine algorithms (Alan et. al., 2015). I4.0 also utilizes mobile technologies to improve connectivity and transparency in daily industrial operations. I4.0 provides technology-based solutions that are environmentally friendly and profitable. The logistics sector focuses on environmental goals that can be achieved at a reasonable cost without compromising the profitability and scalability of the I4.0 technologies.

[a]vijay.bml2017@gmail.com

DOI: 10.1201/9781003450917-1

I4.0 is a well-established term globally with a business of 114.55 billion USD in the year 2021 (Industry 4.0 Market Size, 2022). Everything in the supply chain becomes "smart" due to I4.0 technologies from production to warehouses and logistics. It has shown an uprising trend throughout the world for complete automation. In 2021, the global market for warehousing facilities comprises a value of US$ 451.9 billion. It is expected to rise to US$ 605.6 billion by 2027 as per IMARC Group (Warehousing and Storage Market, 2022). Every business owner needs warehouses and storage facilities for an efficient and seamless setup of the inventory.

Warehouse is a location for collecting and storing raw materials, processed goods, other commodities, etc. The products must be kept in storage so they may be made available to end users as needed. Every step in the value chain of any product requires the storage of a specific quantity of products (van den Berg and Zijm, 1999). Marketing success depends on making the proper arrangements to sell the products in the best possible shape. An organization may continue manufacturing in anticipation of future demand thanks to storage.

A smart warehouse is a warehouse where various components of warehousing operations are automated for improved productivity, efficiency, and accuracy. Warehousing is undergoing a tech-driven revolution as more and more companies adopt smart technologies to cut costs, optimize operations, and improve the overall efficiency of logistic networks (Kembro and Norrman, 2022). I4.0 create warehouse facilities with self-aware, connected items that can exchange data about their location, usage pattern, storage conditions, and more. Everything from product quality assurance and customer service to logistics is managed by smart warehouses. They may also foresee service requirements, get upgrades remotely, and pave the way for fresh, service-based business models for an effective warehouse management system (WMS) (Karpova, 2022). Smart warehouses provide certain advantages over traditional warehouses with lower cost and swift shipping. I4.0 technology integration provides better use of IT infrastructure for the logistics for having a transparent and accurate delivery of product for smoother and precise last mile delivery (LMD) operations. In smart warehouse

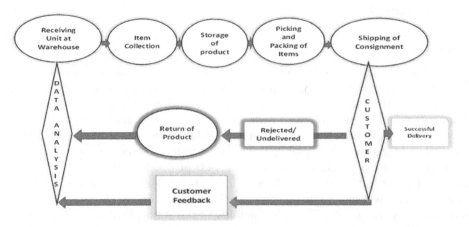

Figure 1.1 Workflow for a traditional warehouse unit.

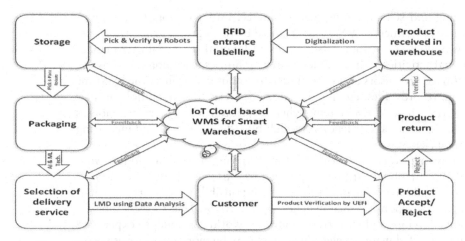

Figure 1.2 Workflow of IoT cloud-based WMS enabled smart warehouse.

the data analysis and forecasting are done with big data analysis and machine learning for managing the daily and seasonal demands. It tries to reduce the bullwhip effect leading to cost expenditures and delays at LMD.

WMS can monitor the warehouse's daily operational activities and try to search for improvement by AI-based machine learning to make supply chain distribution more efficient. WMS systems aid in determining any issues in warehouse operations though they can collect real-time data and create visual reports. WMS data is graphically represented in a simple interface for access to customizable reports and dashboards (Chackelson et. al., 2013). After seeing a WMS report, one could take the proper measures to rectify any anomalies and restart operations. The key benefit of the WMS is that it handles supply chain components other than inventory and order management, including labor management, financial management, reporting, and more. It connects the different data sets and retrieves raw data to gain visibility into shipping carriers, tasks, shipments, work kinds, client job volumes, order ageing, warehouse space, inventory levels, order status, and more. WMSs have some sort of labor-management feature that is useful for scheduling. Sometimes extra workers are required at the ports to swiftly unload new shipments and prevent merchandise from overstocking dock areas, and other times a large order has to be immediately gathered and packaged (Souza et. al., 2022). Demand for and allocation of labor is significantly impacted by seasonal peaks and falls. Automated schedulers aid in forecasting and planning the precise number of workers required by day, zone, and job type by your shipping and procurement schedule (Mirzaei et. al., 2021). The implementation of a WMS is the cherry on top of the smart warehouse technology. WMS system tracks warehouse's everyday operations and determine the methods to improve it by self-learning by technologies like artificial intelligence and machine learning by adopting I4.0 supply chain mechanism. WMS in smart warehouses helps to expose any weaknesses in warehouse operations and tries to provide the efficient processing. WMS reports helps to take the required steps to rectify issues and continue operations. This makes the smart warehouses self-optimizing and self-sustaining. Unicommerce, Luminate logistics, Logiwa, Sortly etc. are the frequently WMS systems used by the logistics firms. These systems help warehouse to maintain the records and digitalize the data for managing operational activities at different levels in warehouses using image processing technique. Image processors and optical readers are used to collect the real-time physical data, records, files, etc. for digital use for real time operations in I4.0 governed WMS. These WMS systems when integrated with smart warehouses can make the operational efficiency very high by reducing errors and increasing operational efficiency.

Customer awareness has increased a lot in the recent decade due to the availability of all information regarding the product by different resources. Conscious customers demand environmental and social responsibility from firms, including sustainable warehousing (Schumacher et. al., 2016). Due to technological advancements develops and industrialization, the world continues to experience the consequences of climate change. Customers may be more loyal to businesses that practice sustainability in operations by paying attention to natural and human resources. Green logistics integration into storage as a component of the supply chain may potentially increase company efficiency and boost financial performance. Applying green trends to warehousing gives a quick return on investment as well as longer-term gains for the world while meeting stakeholder needs (Zhang et. al., 2020). WMS makes it possible to coordinate shipping and transportation while also digitally managing retail, distribution hubs, and warehouses. Green warehouses can run with the right quantity of merchandise thanks to improved inventory control provided by a WMS, minimizing superfluous inventory and maximizing storage space (Komninos, 2022).

It is challenging to put green measures into place throughout the supply chain. But with the correct tools and managerial techniques, businesses can handle a variety of difficulties. It is complicated to manage people and equipment on a wide scale because of the size of the global supply chain. A vast range of transportation kinds, routes, sizes, configurations, as well as humanitarian and social considerations are all covered by supply chain logistics (Mangla et. al., 2016). Compared to other areas of the supply chain, including transportation, storage has less environmentally friendly options. But numerous ways work, and the number of options is expanding. Green logistics' initial outlay, meanwhile, frequently pays dividends over time. This study focuses on the need to investigate I4.0 solutions for modern warehouse management. It explores and analyses the key performance indicators for sustainability (Bank and Murphy, 2013).

Literature background

There have been four industrial revolutions since the 1800s. Each was propelled by a revolutionary new technology: the assembly line's ingenuity, the steam engine's physics, and the computer's speed. They were referred to as industrial "revolutions" because their invention significantly transformed design, manufacturing, production and delivery processes. With each industrial revolution, there has been a significant change in worker skillset requirements (Wang et. al., 2016). The boundaries between the physical, digital, and biological realms is blurring as a consequence of technological convergence. Rate at which technology advancements have no precedent in the past. In contrast to past versions, I4.0 is developing exponentially rather than linearly. Additionally, it is impacting practically every industry worldwide. The size and complexity of these changes also indicate a comprehensive reorganizational need for administration, control, and production systems.

A. Logistics 4.0 and WMS

The usage and development internet-based technologies in the industry have grown inevitable during the past ten years. A new set of logistical challenges brought about by the IoT may require technological advancements, including a critical need for supply chain visibility and integrity regulator (that is right products, at the right time, place, quantity condition, and at the right cost) in supply chains. These changes establish Logistics 4.0 as a concept (Radivojević and Milosavljević, 2019). Warehouses act as the backbone for any logistics firm for smooth operations. In the era of I4.0, there is a need for a sustainable warehouse management system to achieve business goals.

WMSs help in monitoring important warehouse functions such as the number of products selected, the number of orders packaged, travel time, etc. It keeps management informed, awards more productive employees and identifies those not reaching standards or in need of further instruction (Khan et. al., 2022). Additionally, access to operational data and peer comparisons, as well as visibility of those results, boost employee engagement. The adoption of WMS is a good choice for inventory management and production management since it can handle essential company operations from a single system. It was selected as a top option for distribution and manufacturing businesses. Including a method for tracking perishables that entail keeping track of expiration dates and shelf life as well as giving particular goods priority. It provides all the capabilities required to simplify warehouse operations and guarantee on-time delivery. It has a significant effect on improving the bottleneck of the delivery system which is LMD (Sorooshian et. al., 2022).

B. KPIs for WMS in logistics 4.0

a) *Adoption of IOT-based RFID system* (Khan et. al., 2022; Attaran, 2020; Tu et. al., 2018): IoT refers to a variety of digitally connected devices that interact to exchange data with each other. Intelligent warehouse systems emphasizing on all the technical aspects that robots need to link for making warehouse operations efficient. Warehouse management system (WMS) can determine the precise quantity and frequency of items by using an RFID scanner to read the tags. RFID assist in inventory management and organization All operations are automated and data is secured with effective data base management.

b) *Automated guided vehicles* (Radivojević and Milosavljević, 2019; Mirzaei et. al., 2021; Bachofner et. al., 2022): AGVs, which frequently take the place of forklifts, advance material handling and freight transportation. In addition, warehouse robots take care of product picking and packaging. They are essentially more mobile, automated pallet jacks. They transport more stuff at once, more quickly than people, and even figure out the best path to take to collect the required commodities.

c) *AI-based machine learning strategies* (Ding et. al., 2021; Kawa et. al., 2018; Mangiaracina et. al., 2019): AI is becoming an increasingly prevalent technology in all sectors and businesses. The capacity of AI to increase production with few mistakes is the key justification. Robots in warehouses may choose things more effectively by utilizing AI to determine the best path. The kind, quantity, size, and weight of the items may be utilized to identify the appropriate box type for a shipment. Some warehouses are even able to deploy packing devices that use AI to pack goods in the most effective way possible.

d) *Mobile communication-based warehouse mobility* (Ding et. al., 2021; Mao et. al., 2018; Varghese and Tandur, 2014): Employee productivity is increased by the ability to access data while working on the go thanks to mobile devices and applications. Employees are escaping the restrictions of a PC thanks to smartphones. In order to speed up the administration of commodities, they are introducing additional functions and processing capacity to the warehouse systems. Images and video calls may be used to assist people to for transferring information about the products. Quality checks and approvals can be also carried out via a video conference call.

e) *Delivery drones for LMD* (Ren et. al., 2022; Lemardelé et. al., 2021; Ambatkar and Dhatrak, 2022): Drones are employed to provide safe and convenient access to remote, difficult locations. Drones are faster and more accurate than manual processes since they are equipped with cameras, sensors, RFID, or barcode scanners. They are used, among other inventory-related tasks, to locate items and perform cycle counts.

C. Four pillars of sustainability

a) *Economic development*: Transportation, storage, and distribution are all part of the logistics process, which seeks to fulfil consumer demand while optimizing product profitability from source point to destined location (Chin et. al., 2008). WMS has grown in importance as a means for industries to boost their accuracy and logistics cycle efficiency. It has emerged as a key factor and predictor for combining global value chains and the maintenance activities of economic development (Van Geest et. al., 2022).

b) *Humanitarian aspect*: Humanitarian aspects covers a variety of people with independent viewpoints. A number of variables, including different technologies, skills, human emotions and managerial operations contribute to the complexity of the humanitarian system. Additionally, the type, size, strategy, goal, specialty, rules and regulations, and work culture improvement scope also impacts the workers and staff for creating a significant change (Jiang and Yuan, 2019). Therefore, the humanitarian aspect remains an important factor to achieve sustainability.

c) *Social consequences*: Managing the good and negative effects of the I4.0 digital revolution on people is one of the key aspects of social implications.

The involvement of the employees and staff as shareholders is critical for any organizational upgradation (Tsarouchi et. al., 2016). Companies either directly or indirectly have an impact on the lives of their workers, value chain members, customers, and community groups (Matthess et. al., 2022).

d) *Environmental concerns*: Global supply networks' logistical operations are becoming a significant source of industrial pollutants for causing environmental damage (Toktać-Palut, 2022). Although storage and product management procedures in warehouses are substantial source carbon emissions. I4.0 helps to discover numerous methods for lowering the carbon footprint and warehouse emissions to improve environmental quality (Kinkel et. al., 2022).

Methodology

One of the most well-known and often used multicriteria procedures is the analytical hierarchy process (AHP) (Bakhtari et. al., 2021; Wang et. al., 2008). This approach combines the steps of evaluating options and groups them to select the most important ones. This technique is used to rank a group of options or to choose the ideal option from a list of possibilities. Rankings and choices are determined in view of a broad objective that is broken down into a number of components (Qureshi, 2022). AHP is found suitable to determine the importance of investigated I4.0 KPIs from the literature and expert discussion for all four sustainability criteria. The priority order and importance of I4.0 KPIs are determined for sustainable WMS.

A. Implementation of AHP

AHP provides the nomenclature to I4.0 KPIs for WMS and sustainability evaluation criteria in Tables 1.1 and 1.2. The weightage is assigned as per the data collected in the survey on a scale of 1.1–1.5. Table 1.3 calculates the eigen vector, weight and comp eigen vector for sustainability criteria. Tables 1.4–1.7 presents KPIs evaluation with respect to each sustainability evaluation criteria.

Results

It is very evident from the implementation of AHP analysis on I4.0 technologies that economic development (CRI1) and humanitarian aspects (CRI2) are most impacted by AI-based machine learning strategies (KPI3). will be the most

Table 1.1: I4.0 KPIs for WMS

S.No.	KPIs	Abbreviation/Code
1	Adoption of IoT-based RFID system	KPI1
2	Automated guided vehicles	KPI2
3	AI-based ML strategies	KPI3
4	Mobile communication-based warehouse mobility	KPI4
5	Delivery drones for LMD	KPI5

Table 1.2: Sustainability evaluation criteria

S.No.	Criteria	Abbreviation/Code
1	Economic development	CRI1
2	Humanitarian aspect	CRI2
3	Social consequences	CRI3
4	Environmental concerns	CRI4

Table 1.3: KPIs with respect to goal sustainability logistics

	CRI1	CRI2	CRI3	CRI4	Eigen vector	Weight	Comp eigen vector
CRI1	1	5	3	3	2.59	0.5409	2.1782
CRI2	1/5	1	1/4	3	0.6223	0.13	0.7568
CRI3	1/3	4	1	1/3	0.8165	0.1705	0.9235
CRI4	1/3	1/3	3	1	0.7598	0.1587	0.8938

λ_{max} = 4.7523, Consistency Index = 0.2508, Consistency Ratio =0.2786

Table 1.4: KPIs with respect to the economic development

	KPI1	KPI2	KPI3	KPI4	KPI5	Eigen vector	Weight	Comp eigen vector
KPI1	1	1/3	3	1/5	1/4	0.5493	0.0998	1.1616
KPI2	3	1	1/3	4	3	1.6438	0.2985	1.7862
KPI3	1/3	3	1	4	3	1.6438	0.2985	2.3163
KPI4	5	1/4	1/4	1	3	0.9872	0.1793	1.1991
KPI5	4	1/3	1/3	1/3	1	0.6826	0.124	0.7817

λ_{max} =7.245, Consistency Index = 0.5612, Consistency Ratio =0.5011

Table 1.5: KPIs with respect to humanitarian aspect

	KPI1	KPI2	KPI3	KPI4	KPI5	Eigen vector	Weight	Comp eigen vector
KPI1	1	1/3	3	1/5	1/4	0.5493	0.0998	1.1616
KPI2	3	1	1/3	4	3	1.6438	0.2985	1.7862
KPI3	1/3	3	1	4	3	1.6438	0.2985	2.3163
KPI4	5	1/4	1/4	1	3	0.9872	0.1793	1.1991
KPI5	4	1/3	1/3	1/3	1	0.6826	0.124	0.7817

λ_{max} =6.5271, Consistency Index = 0.3818, Consistency Ratio =0.3409

Table 1.6: KPIs concerning social consequences

	KPI1	KPI2	KPI3	KPI4	KPI5	Eigen vector	Weight	Comp eigen vector
KPI1	1	5	4	3	1/3	1.8206	0.3166	1.7734
KPI2	1/5	1	3	1/4	1/3	0.5493	0.0955	0.642
KPI3	1/4	1/3	1	4	1/3	0.6444	0.1121	0.8935
KPI4	1/3	4	1/4	1	1	0.8027	0.1396	0.9914
KPI5	3	3	3	1	1	1.9332	0.3362	2.0484

λ_{max} =6.3488, Consistency Index = 0.3372, Consistency Ratio =0.3011

Table 1.7: KPIs for environmental concerns

	KPI1	KPI2	KPI3	KPI4	KPI5	Eigen vector	Weight	Comp eigen vector
KPI1	1	3	1/3	4	3	1.6438	0.2792	1.9268
KPI2	1/3	1	1/5	1/3	3	0.5818	0.0988	0.5378
KPI3	3	5	1	1/4	4	1.7188	0.2919	1.9556
KPI4	1/4	3	4	1	3	1.5518	0.2636	1.9971
KPI5	1/3	1/3	1/4	1/3	1	0.392	0.0666	0.3534

λ_{max} =6.7708, Consistency Index = 0.4427, Consistency Ratio =0.3953

Table 1.8: Final priority order

S.No.	KPIs	Priority values
1	Adoption of IoT-based RFID system	0.2695
2	Automated guided vehicles	0.1207
3	AI-based machine learning strategies	0.2518
4	Mobile communication-based warehouse mobility	0.2384
5	Delivery drones for LMD	0.1196

prominent. Delivery drones for LMD (KPI5) are the most prominent I4.0 technologies concerning social consequences (CRI3). Mobile communication-based warehouse mobility (KPI4) plays an important role in environmental concerns (CRI4). The final priority order for KPIs to achieve a sustainable WMS system is shown in Table 1.8.

Conclusion

The study of I4.0 technology adoption for sustainable warehouse management considers the following priority order.

KPI1>KPI3>KPI4>KPI2>KPI5

Adoption of an IoT-based RFID system that automates or digitizes order picking using improves accuracy and expedites warehouse orders. The use of less paper is an extra benefit of IoT-based order-picking technology. It has proved to be the most significant KPI for the sustainability of the warehouse. The environmentally friendly distribution techniques of barcoding and RFID also lessen the demand for paper.

Through the supply chain, these technologies enhance procedures and boost product visibility. Any warehouse operation may be significantly optimized by employing smart warehousing. AI-based machine learning technologies are developing an ecosystem of "smart" warehouses that offer visibility and give supply chains the agility, flexibility, and reactivity they require with much dependence on human interventions. Delivery drones seem to be the less important part of sustainable WMS concerning other KPIs. They are independently providing services on cloud-based computing for improving the LMD.

Future scope

The majority of the process, from suppliers to customers, is handled by the new smart warehouse systems. As automation becomes more prevalent, warehouse companies will be able to use it more consistently. Automated guided vehicle technology is less time-consuming and reduces costs while eliminating the inherent risks of human effort in the movement of goods. A separate case study can be done to determine sustainable warehouse efficiency. The absence of standards across relief groups was one of the greatest issues that logisticians working in humanitarian aid encountered throughout the debate. A standardized humanitarian scale based on labor rights and human ethics can be made for all smart warehouses.

References

Ambatkar, H. P. and Dhatrak, R. K. (2022). Review: drone applications in transmission line. *In International Mobile and Embedded Technology Conference, MECON,* (2022), (pp. 531–533). https://doi.org/10.1109/MECON53876.2022.9752257.

Attaran, M. (2020). Digital technology enablers and their implications for supply chain management. *Supply Chain Forum*, 21(3), 158–172. https://doi.org/10.1080/16258 312.2020.1751568.

Bachofner, M., Lemardelé, C., Estrada, M., and Pagès, L. (2022). City logistics: challenges and opportunities for technology providers. *Journal of Urban Mobility*, 2, 100020. https://doi.org/10.1016/J.URBMOB.2022.100020.

Bakhtari, A. R., Waris, M. M., Sanin, C., and Szczerbicki, E. (2021). Evaluating industry 4.0 implementation challenges using interpretive structural modeling and fuzzy analytic hierarchy process. *Cybernetics and Systems*, 52(5), 350–378. https://doi.org/10 .1080/01969722.2020.1871226.

Bank, R. and Murphy, R. (2013). Warehousing sustainability standards development. *IFIP Advances in Information and Communication Technology*, 414, 294–301. https://doi.org/10.1007/978-3-642-41266-0_36.

Chackelson, C., Errasti, A., Ciprés, D., and Lahoz, F. (2013). Evaluating order picking performance trade-offs by configuring main operating strategies in a retail distributor: a design of experiments approach. *International Journal of Production. Research*, 51(20), 6097–6109. https://doi.org/10.1080/00207543.2013.796421.

Chin, K. S., Chan, B. L., and Lam, P. K. (2008). Identifying and prioritizing critical success factors for coopetition strategy. *Industrial Management and Data Systems*, 108(4), 437–454. https://doi.org/10.1108/02635570810868326.

Ding, Y ., Jin, M., Li, S., and Feng, D. (2021). Smart logistics based on the internet of things technology: an overview. *International Journal of Logistics Research and Applications*, 24(4), 323–345. https://doi.org/10.1080/13675567.2020.1757053

Industry 4.0 Market Size, Share | Growth Analysis [2022-2029]. (n.d.). Industry 4.0 Market Size, Share | Growth Analysis [2022-2029]. https://www.fortunebusinessinsights.com/industry-4-0-market-102375.

Jiang, Y. and Yuan, Y. (2019). Emergency logistics in a large-scale disaster context: achievements and challenges. *International Journal of Environmental Research and Public Health*, 16(5), 779. https://doi.org/10.3390/ijerph16050779.

Karpova, N. P. (2022). Modern warehouse management systems. *Lecture Notes in Networks and Systems*, 304, 261–267. https://doi.org/10.1007/978-3-030-83175-2_34.

Kawa, A., Pieranski, B., and Zdrenka, W. (2018). Dynamic configuration of same-day delivery in e-commerce. *Studies in Computational Intelligence*, 769, 305–315. https://doi.org/10.1007/978-3-319-76081-0_26.

Kembro, J. and Norrman, A. (2022). The transformation from manual to smart warehousing: an exploratory study with Swedish retailers. *International Journal of Logistics Management*, 33(5), 107–135. https://doi.org/10.1108/IJLM-11-2021-0525.

Khan, M. G., Ul Huda, N., and Uz Zaman, U. K. (2022). Smart warehouse management system: architecture, real-time implementation and prototype design. *Machines*, 10(2), 150. https://doi.org/10.3390/MACHINES10020150.

Kinkel, S., Baumgartner, M., and Cherubini, E. (2022). Prerequisites for the adoption of AI technologies in manufacturing – evidence from a worldwide sample of manufacturing companies. *Technovation*, 110, 102375. https://doi.org/10.1016/j.technovation.2021.102375.

Komninos, N. (2022). Transformation of industry ecosystems in cities and regions: a generic pathway for smart and green transition. *Sustainability*, 14(15), 9694. https://doi.org/10.3390/SU14159694.

Lemardelé, C., Estrada, M., Pagès, L., and Bachofner, M. (2021). Potentialities of drones and ground autonomous delivery devices for last-mile logistics. Transportation Research Part E: Logistics and Transportation Review, 149(C), 305–315. https//doi.org/10.1016/j.tre.2021.102325.

Mangiaracina, R., Perego, A., Seghezzi, A., and Tumino, A. (2019). Innovative solutions to increase last-mile delivery efficiency in B2C e-commerce: a literature review. *International Journal of Physical Distribution and Logistics Management*, 49(9), 901–920. https://doi.org/10.1108/IJPDLM-02-2019-0048.

Mangla, S. K., Govindan, K., and Luthra, S. (2016). Critical success factors for reverse logistics in Indian industries: a structural model. *Journal of Cleaner Production*, 129, 608–621. https://doi.org/10.1016/J.JCLEPRO.2016.03.124.

Mao, J., Xing, H., and Zhang, X. (2018). Design of intelligent warehouse management system. *Wireless Personal Communications*, 102(2), 1355–1367. https://doi.org/10.1007/s11277-017-5199-7.

Matthess, M., Kunkel, S., Xue, B., and Beier, G. (2022). Supplier sustainability assessment in the age of industry 4.0 – insights from the electronics industry. *Cleaner Logistics and Supply Chain*, 4(100038), 1–19. https://doi.org/10.1016/J.CLSCN.2022.100038.

Mirzaei, M., Zaerpour, N., and de Koster, R. (2021). The impact of integrated cluster-based storage allocation on parts-to-picker warehouse performance. *Transportation Research Part E: Logistics and Transportation Review*, 146, 102–207. https://doi.org/10.1016/j.tre.2020.102207.

Qureshi, M. R. N. M. (2022). A bibliometric analysis of third-party logistics services providers (3PLSP) selection for supply chain strategic advantage. *Sustainability*, 14(19), 11836. https://doi.org/10.3390/su141911836.

Radivojević, G. and Milosavljević, L. (2019). The concept of logistics 4.0. *In 4th Logistics International Conference*, (pp. 283–293). https://logic.sf.bg.ac.rs/wpcontent/uploads/LOGIC_2019_ID_32.pdf.

Ren, X., Vashisht, S., Aujla, G. S., and Zhang, P. (2022). Drone-edge coalesce for energy-aware and sustainable service delivery for smart city applications. *Sustainable Cities and Society*, 77, 103–505 . https://doi.org/10.1016/j.scs.2021.103505.

Saldivar, A. A. F., Li, Y., Chen, W. N., Zhan, Z. H., Zhang, J., & Chen, L. Y. (2015, September). Industry 4.0 with cyber-physical integration: A design and manufacture perspective. In 2015 21st international conference on automation and computing (ICAC) (pp. 1-6). IEEE. https://doi.org/10.1109/IConAC.2015.7313954.

Schumacher, A., Erol, S., and Sihn, W. (2016). A maturity model for assessing industry 4.0 readiness and maturity of manufacturing enterprises. *Procedia CIRP*, 52, 161–166. https://doi.org/10.1016/j.procir.2016.07.040.

Sorooshian, S., Khademi Sharifabad, S., Parsaee, M., and Afshari, A. R. (2022). Toward a modern last-mile delivery: consequences and obstacles of intelligent technology. *Applied System Innovation*, 5(4), 82. https://doi.org/10.3390/asi5040082.

Souza, F. F. de, Corsi, A., Pagani, R. N., Balbinotti, G., and Kovaleski, J. L. (2022). Total quality management 4.0: adapting quality management to industry 4.0. *TQM Journal*, 34(4), 749–769. https://doi.org/10.1108/TQM-10-2020-0238.

Toktaş-Palut, P. (2022). Analyzing the effects of industry 4.0 technologies and coordination on the sustainability of supply chains. *Sustainable Production and Consumption*, 30, 341–358. https://doi.org/10.1016/J.SPC.2021.12.005.

Tsarouchi, P., Makris, S., and Chryssolouris, G. (2016). Human–robot interaction review and challenges on task planning and programming. *International Journal of Computer Integrated Manufacturing*, 29(8), 916–931. https://doi.org/10.1080/0951 192X.2015.1130251.

Tu, M., Lim, M. K., and Yang, M. F. (2018). IoT-based production logistics and supply chain system–Part 2: IoT-based cyber-physical system: a framework and evaluation. *Industrial Management and Data Systems*, 118(1), 96–125. https://doi.org/10.1108/ IMDS-11-2016-0504.

van den Berg, J. P., and Zijm, W. H. M. (1999). Models for warehouse management: classification and examples. *International Journal of Production Economics*, 59(1), 519–528. https://doi.org/10.1016/S0925-5273(98)00114-5.

Van Geest, M., Tekinerdogan, B., and Catal, C. (2022). Smart warehouses: rationale, challenges and solution directions. *Applied Sciences (Switzerland)*, 12(1), 1–16. https:// doi.org/10.3390/app12010219.

Varghese, A. and Tandur, D. (2014). Wireless requirements and challenges in industry 4.0. *In International Conference on Contemporary Computing and Informatics (IC3I)*, *Mysore, India*, 2014, (pp. 634–638). https://doi.org/ 10.1109/IC3I.2014.7019732.

Wang, S., Wan, J., Zhang, D., Li, D., and Zhang, C. (2016). Towards smart factory for industry 4 . 0 : a self-organized multi-agent system with big data based feedback and coordination. *Computer Networks*, 101, 158–168. https://doi.org/10.1016/j. comnet.2015.12.017.

Wang, Y. M., Luo, Y., and Hua, Z. (2008). On the extent analysis method for fuzzy AHP and its applications. *European Journal of Operational Research*, 186(2), 735–747. https://doi.org/10.1016/J.EJOR.2007.01.050.

Warehousing and Storage Market (2022). Warehousing and storage market: global industry trends, share, size, growth, opportunity and forecast 2022-2027. Imarcgroup. https://www.imarcgroup.com/warehousing-and-storage-market#:~:text=The global warehousing and storage,4.9%25 during 2022-2027.

Zhang, W., Zhang, M., Zhang, W., Zhou, Q., and Zhang, X. (2020). What influences the effectiveness of green logistics policies? a grounded theory analysis. *Science of the Total Environment*, 714, 136731. https://doi.org/10.1016/j.scitotenv.2020.1367.

2 An optimal load frequency controller employing particle swarm optimization algorithm for an autonomous hybrid renewable energy system

C. Suchetha[a] and J. Ramprabhakar[b]

Department of Electrical and Electronics Engineering, Amrita School of Engineering, Bengaluru, Amrita Vishwa Vidyapeetham, India

Abstract

An optimized load frequency controller employing an evolutionary algorithm is proposed for an autonomous hybrid renewable energy system. The primary goal is to maintain load voltage and frequency while working independently. Conventional PI regulators are used for the control of power and current. The PI regulators are tuned for optimal values by implementing an evolutionary search algorithm, "particle swarm optimization algorithm". The distinctive aspect of this approach is that the global optimum obtained stays optimal for all possible input variable values that are within operational limits. This method eliminates the need for a real-time search algorithm. The proposed method is justified by the results obtained.

Keywordshy: Brid power controller, hybrid renewable energy system, load frequency control, particle swarm optimization, Hybrid power controller

Introduction

Optimization is defined as the best and effective use of available resources. Looking at the present energy scenario, alternative cum renewable energy sources provide a propitious solution. The catch here is the unpredictable nature and non-transportable nature of renewable energy sources and, optimization looks like the only option for sustenance. To overcome, the stochastic nature of renewable energy sources, to extract maximum power from natural resources, it is an accepted practice to interconnect 2 or more distributed energy sources to form a micro-grid The infrastructure consists of maximum power trackers, to extract the maximum available power and a load/grid end power converter to deliver better quality power. The incidence of harmonics that degrade the quality of power rises with the adoption of power electronic circuits.

To maintain the quality of power, it is very important to tune controller parameters properly. But, the unreliable nature of the energy sources makes it a very difficult task. Different researchers have implemented various methods to subdue this issue; the most popular trends are reported in (Suchetha and Ramprabhakar, 2018). An islanded microgrid with a 3LNCI employed the droop control method and complications in this method are overcome by using the virtual resistor

[a]c_suchetha@blr.amrita.edu, [b]j_ramprabhakar@blr.amrita.edu

DOI: 10.1201/9781003450917-2

(VR) active damping method (Gervasio et. al., 2015). Another method of droop control is demonstrated by Ramezani and Li, (2016) by combining it with the current vector control method. The DC bus voltage is stabilized by using an ultra-capacitor for faster response (Jamshidpour et. al., 2015). A different method of controlling load voltage is demonstrated by implementing a FACTS-based compensator for the power filter switched by PWM (Sharaf et. al., 2007). Load voltage and frequency are controlled using an advanced VSI controller (Alam et. al., 2013). Non-linear controllers that are robust in nature are connected at each distributed energy resource (Hoseini et. al., 2020). Advanced control tactics are employed to achieve global stability.

Optimization techniques, when implemented will be able to tune controller parameters that can deliver optimal power. Literature shows various methodologies of applying optimization methods for improving power quality. A distributed cooperative tracker algorithm is implemented by Dehghan Banadaki et. al (2017) to control terminal voltage in a microgrid operating autonomously. A simulated annealing method is employed (Chandrakala and Balamurugan, 2016) to accomplish automated voltage and load frequency control in a multi-area multi-source system. Whereas (Mohanty et. al., 2014). differential evolution algorithm is employed to calculate the variables of the load frequency controller. A load frequency controller (LFC) supported by electric vehicles (EV) is employed for an islanded microgrid (Khooban et. al., 2018). The LFC comprises of "fractional-order fuzzy PID (FOFPID) controller" whose parameters are automatically optimized by deploying a "modified black-hole" approach. Optimal generation and voltage regulation in a microgrid is achieved (Mudaliyar et. al., 2020) by using distributed economic dispatch. Active power balance is achieved by an electronic load controller (Serban and Marinescu, 2011), this combines the usage of BESS and a smart-load. A three-area deregulated load frequency controller (Lekshmi et. al., 2016) employs PI controllers which are optimally tuned using fuzzy gain scheduling. The suitability of direct power control for a bi-directional is evaluated (Warrier et. al., 2020). Model prediction control acts as an add-on in predicting the reference current and improves dynamic response of the system. An isolated power system containing renewable energy sources has its load frequency managed using model predictive control (Liu et. al., 2020).

Two control loops are employed (Karimi et. al., 2016), the outer loop directly regulates terminal voltage and frequency, whereas the inner loop controls voltage and current. Under fluctuating load conditions, for the system to operate optimally the variables are tuned using particle swarm optimization (PSO). VSI controller, dynamic filter and green power filter (Sharaf and El-Gammal, 2009) are equipped with PI controllers, whose parameters are tuned by implementing PSO algorithm. A power controller with online tuning is commissioned (Al-Saedi et. al., 2017), PSO algorithm is used for tuning parameters. For a grid-connected system (Al-Saedi et. al., 2013) a controller is auto-tuned using PSO. The control performance and power quality are maintained (Chung et. al., 2008) by tuning the controller with PSO algorithm. Load frequency control, along with other objectives are solved (Pandi et. al., 2017) using fuzzy adapted PI controller, which in turn is optimized by PSO. A variation of PSO that combines the search process of a differential optimization algorithm is employed along with mixed integer programming (Duman et. al., 2020) to regulate the flow of power in a distributed power generating system with solar and wind as sources. Power flow controller is proposed (Suchetha and Ramprabhakar, 2019; Jumani et. al., 2018) for optimal sharing of power between the local load and main grid. The controller consists of proportional-integral controller variables are tuned using grasshopper optimization algorithm (GOA) (Jumani et. al., 2018) and PSO (Suchetha and Ramprabhakar, 2019). Many researchers have explored the topic of employing soft computing methods for improved operation of the system. Form the reported results, where problems that are unconstrained and continuous, it is evident that majority of methods applied have some shortcomings. PSO on the other hand is more efficient in many ways like speed of calculating and utilizing the search area as reported (Li et. al., 2017; Jumani et. al., 2018). The interdependency of line frequency and terminal voltage is exploited using game theory (Younesi et. al., 2020). A new objective is optimized (Barik and Das, 2020). Integral square of weighted absolute error (ISAE) is considered as an objective and solved using quasi-oppositional selfish herd optimization. In order to increase the performance of the controller, the gain parameters were adjusted using the PSO approach with the aid of four cost functions (Dhanasekaran et. al., 2023).

The results indicate that the PSO-PID controller results in a quicker response time and the proposed method shows a 79% improvement over the traditional method. The load-frequency controlling issue of a two-area linked electric power system is addressed using a unique cascaded "3DOF-FOPID-FOPI" controller (Xie et. al., 2023). The LFC of two-area non-reheat thermal systems using a unique artificial rabbits algorithm (ARA) is illustrated in (El-Sehiemy et. al., 2023).

The present work proposes a system that controls the operation of VSI in autonomous mode. The controller generates reference voltages in synchronous d-q frame which are interfaced by a "space vector pulse width modulation SVPWM" to produce control signals for VSI. High dynamic response is achieved by implementing a voltage feed forward compenzation in the current loop. Integral time average error (ITAE) solved using Simpson's 1/3 rule is considered as the objective for population based optimization algorithm. Real time online optimization, even though attempted is difficult to implement because, of various reasons viz., searching process is lengthy, more simulation time and continuously changing load.

To demonstrate the performance of the suggested controller, simulation is run. The study's primary contributions are listed below.

1. A 2-stage controller is adapted to manage voltage and power fed to load.
2. PSO evolutionary algorithm is implemented to optimally calculate the parameter values.
3. A unique search method is proposed whose performance is in par with a real-time search algorithm.
4. The optimization is carried out using a set of input parameters that are most likely to occur in the specific location, ensuring the validity of the optimized values under various operating conditions.

The proposed controller is simulated in both autonomous mode and when connected to main grid mode and its effectiveness is tested for variable solar irradiation, wind speed.

This article is organized further into four sections. First system layout and mathematical models of each component is described. The controller design and mathematical conversions are discussed in the next part. The definition of PSO and its appropriateness and compliance with the present problem is described in section four. Also, the objective function is defined in this section. Different scenarios to which the system is subjected and the results obtained are discussed in section five.

Model overview

An operating model consisting of an energy storage system and various renewable/alternate energy sources is designed in SIMULINK. The model consisting of various energy sources is termed as a hybrid renewable energy source (HRES). The HRES can operate effectively in both grid-tied mode and disconnected modes. The proposed system has a solar PV system, wind energy conversion system and fuel cell unit as energy sources, and one BESS for backup. The system configuration is graphically represented in Figure 2.1. The configuration information specs are shown in Table 2.1.

A permanent magnet synchronous generator (PMSG) is used in the wind turbine for energy conversion. Since PMSG is a direct drive engine, major drawbacks like: weight of gear-box, losses due to translation, and maintenance requirements are reduced. Linked to PMSG is a rectifier cascaded with a converter. To acquire the preferred voltage level, the solar PV is connected to a boost converter. The energy storage system is configured to offer ancillary assistance during peak load hours. Fuel cell with its small response time it can act as a buffer for surge loads. A voltage source inverter acts as an interface between DC link and load. To

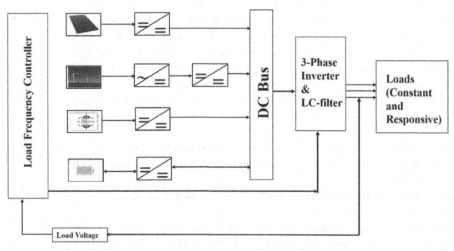

Figure 2.1 Prototype of a hybrid renewable energy system

Table 2.1: Specifications of sources and load

Solar PV	V_{pv}=183V DC	V_{dc} = 380V DC	Power = 60kW
Wind turbine	V_{w} =232V AC	V_{dc} = 380V DC	Power = 28kW
Fuel Cell	V_{fc} =150 V DC	V_{dc} = 380V DC	Power = 6kW
Battery	V_{batt} = 120V DC	V_{dc} = 380V DC	Power = 6kW
Load	P = 96kW	Q = 46kW	S = 100kW

(Source: Author's compilation)

increase the quality of power fed to the load an L-C filter is connected based on the design proposed (Kim et. al., 2011).

Load frequency controller

The presence of multiple and diverse energy sources demand a power-electronic link to ensure the load's power quality. The load voltage and load frequency are interdependent and can be controlled simultaneously. In an autonomous mode of operation, load voltage and frequency specifications are to be met, so a V-F controller controls the inverter.

The proposed V-F control technique consists of two control levels: the level of current control and the level of power control. The loop determines current reference values (I_d, I_q) following equations (2.1). A sluggish rate of variation in reference currents results in improved system control. This implies that controller parameters (K_{vp}) and (K_{vi}) are to be selected accordingly.

$$i_{d^*} = (V^* - V_{load})\left(K_{vp} - \frac{K_{vi}}{s}\right)$$

$$i_{q^*} = (f^* - f_{load})\left(K_{fp} - \frac{K_{fi}}{s}\right) \quad (1)$$

The current control stage is actualized by two controllers similar to level of power control. This step governs the small transient currents at the inverter output. The additional feed-forward voltage loop enhances the dynamic response of the system's and overall stability. Voltage values for reference are generated at this stage in *d-q* reference frame (V_d, V_q). Obtained reference voltages are transformed into gate signals for inverter switches. Mathematical equations for Clark's transformations are displayed in equation (2.2). The space vector pulse width modulation (SVPWM) method is employed to generate six pulses for switches in the inverter. The main feature of SVPWM is that it utilizes maximum input voltage and also provides high harmonic reduction as demonstrated (Syed and Raahemifar, 2014). The voltage reference value is mathematically calculated as equation (2.3). The proposed V-F controller schematic representation is shown in Figure 2.2. The proportional-integral regulator parameters are tuned by PSO.

$$\begin{bmatrix} V_\alpha \\ V_\beta \\ V_0 \end{bmatrix} = \frac{2}{3} \begin{bmatrix} V_a \\ V_b \\ V_c \end{bmatrix} \begin{bmatrix} 1 & \frac{-1}{2} & \frac{-1}{2} \\ 0 & \frac{\sqrt{3}}{2} & \frac{-\sqrt{3}}{2} \\ \frac{1}{2} & \frac{1}{2} & \frac{1}{2} \end{bmatrix} \qquad (2)$$

$$\begin{bmatrix} V_d^* \\ V_q^* \end{bmatrix} = \begin{bmatrix} -K_p & -\omega L_s \\ \omega L_s & -K_p \end{bmatrix} \begin{bmatrix} i_{ld} \\ i_{lq} \end{bmatrix} + \begin{bmatrix} K_p & 0 \\ 0 & K_p \end{bmatrix} \begin{bmatrix} i_d^* \\ i_q^* \end{bmatrix} + \begin{bmatrix} K_i & 0 \\ 0 & K_i \end{bmatrix} \begin{bmatrix} e_d \\ e_q \end{bmatrix} + \begin{bmatrix} V_{sd} \\ V_{sq} \end{bmatrix} \qquad (3)$$

Where, e_d, e_q are the integral current errors.

PSO implementation

Meta-heuristic algorithms are mostly bio-inspired. These algorithms are developed after observing the survivability and hunting of a particular species. By observing a flock of birds and their group behavior, while acquiring food in a space, Kennedy and Eberhart have proposed PSO in 1995 (Eberhart and Kennedy, 1995). Swarm space is defined as the boundaries within which an optimal solution may be discovered. The number of birds searching the space increases the chances of finding solution, similarly more number of particles provide a better solution. After each iteration the best value and position is communicated between all particles. This ensures a faster and more accurate calculation of solution. The algorithm is presented in Figure 2.3. From the figure it is understood that after every iteration the birds move closer to a location called global best position X_{gbest}. Evaluating the algorithm involves three steps viz.

1. Determine the cost of every particle.
2. Compare global best cost position with present best cost and positions.
3. Re-evaluate velocity of each particle to reach next position.

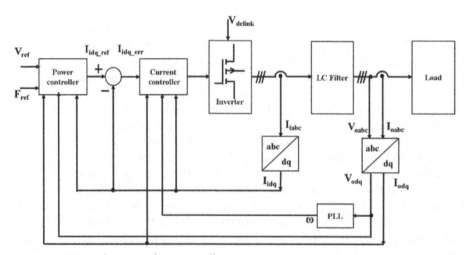

Figure 2.2 Schematic of VF controller

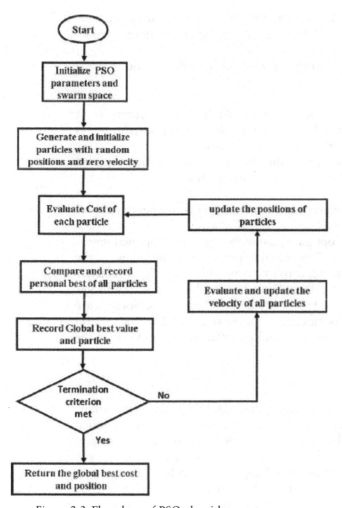

Figure 2.3 Flowchart of PSO algorithm

The velocity of a particle is a function of its previous position and global best cost and local best cost values and their positions respectively. The new position to be reached by each particle is determined from the calculated velocity. Mathematically represented in equations (2.4) and (2.5).

$$V_i^{m+1} = \omega \cdot V_i^m + c_1 \cdot r_1 [X_{pbest}^m - X_i^m] + c_2 \cdot r_2 [X_{gbest}^m - X_i^m] \quad (4)$$

$$X_i^{m+1} = X_i^m + V_i^{m+1} \quad (5)$$

Where i is an index of the particle, m is the number of iteration, w is the inertia constant, within the range of [0, 1], c_1, c_2 are acceleration coefficient, between [0, 2] and, r_1, r_2 are random values. After all the particles are allocated to new positions, next iteration is started in search of the best optimal value. The same actions are repeated until one of the following criteria, the set aim or the number of iterations, is satisfied.

There are many factors that control the performance of the algorithm, some are discussed below:

- i, Number of particles: The population should be chosen optimally as the higher population leads to better solution but, lengthy search process whereas lesser population leads to a local minimum.
- w, Inertia constant: If w is higher, the direction of the next position is more diverse than previous. This leads to a wider search area.

c_1, c_2, acceleration coefficient: Higher values lead to faster convergence but may miss out global best in the process.

The controller parameters are optimized using PSO to obtain optimal operation in all operational conditions. This is highly difficult to achieve due to the stochastic nature of the power sources. To overcome this issue, the parameters are optimized for different input values. These input values are a set of solar irradiations most common for that location and set of wind velocities that are within the cut-in and cut-out velocities and are more frequent. This way, we are avoiding the need for online optimization and maintaining the response of the algorithm.

3.1 Cost function

For a load frequency controller, minimizing voltage and frequency error will be most logical objective. The V/f error is accepted as the objective since terminal voltage and frequency are interdependent. Out of all the error indexes "integral time absolute error (ITAE)" offers the most accurate error. This error index is calculated mathematically by implementing Simpson's 1/3 rule (Bolton, 1997). The numerical articulation for ITAE value is characterized by equation (2.6). From the expression, $e(t)$ is the error that needs to be reduced.

$$\text{ITAE} = \int_0^\infty t|e(t)|\,dt \qquad (6)$$

3.2 Termination criterion

The algorithm will exit the loop in two situations: one is when the defined objective is achieved or, when a number of iterations are completed. For the present application, the algorithm will terminate after the predefined number of iterations are completed. Predefined specifications for running algorithm are listed in Table 2.2.

The algorithm is evaluated in the following steps.

1. Uniform random particles are generated within the search space.
2. Fitness value of each particle is evaluated.

3. Fitness values are compared to find the current global minimum
4. The velocity is each particle is calculated using equation (2.5).
5. The new position of each particle is determined by equation (2.4).

Figure 2.4 shows the heterogeneity of particles in the space moving in the direction of minimum. The PI controller parameters are optimized for voltage, frequency ratio. By simulating each particle in the swarm space and accounting for variations in wind speed and sun irradiation, the PSO algorithm is realized to tweak the constraints. The particles will move toward minimum cost, in our case: integrated time average error. The transition of cost towards global best cost over the iterations is traced in graph and displayed in Figure 2.5. It is evident from the graphs depicting the local minima of the algorithm in Figure 2.5, that the speed of convergence is high. It can be seen that the final value is reached before the number of iterations are completed. The obtained optimum parameters are illustrated in Table 2.3.

Simulation and results

A model of the proposed HRES system is designed in MATLAB/Simulink application. The filter specs are adjusted to filter higher order harmonics, and the DC

Table 2.2: Specifications for algorithm

Particles	50	
Iterations	250	
Initial velocity	0	
Inertia constant	0.7	
Acceleration coefficients	0.1	
Search limit	$Kpv = [0 - 30]$	$Kiv = [0 - 5 * 10^{-3}]$
	$Kpf = [0 - 15]$	$Kif = [0 - 0.001]$

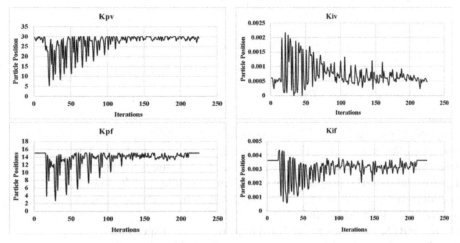

Figure 2.4 Search process of control parameters

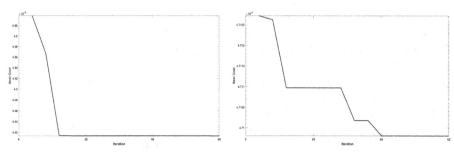

Figure 2.5 Best cost of the parameters

Table 2.3: Power level controller constraints

Control constraints	Best position
K_{pv}	30.00000
K_{iv}	0.000517
K_{pf}	14.10688
K_{if}	0.003146

(Source: Author's compilation)

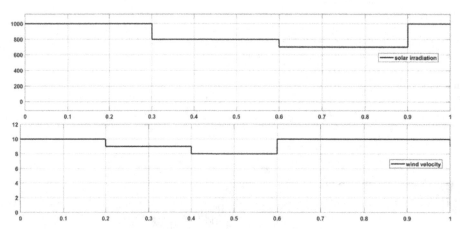

Figure 2.6 Variable input parameters

link capacitor, which maintains the DC voltage level, is set at 5000 µF. The inner current loop controller parameters are tuned using traditional methods. The switching frequency of the SVPWM is set at 20 kHz. Three testing scenarios are conceptualized to validate the proposed model. In these conditions the load is varied from underrated to rated and to above-rated value, the renewable sources vary irrespective of the load connected.

For testing the HRES, the input in the form of variable wind speed and variable solar irradiation is randomly generated, it is indicated in Figure 2.6. The

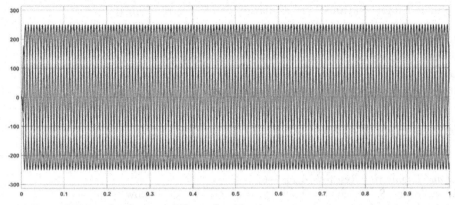

Figure 2.7 Load voltage at different load levels

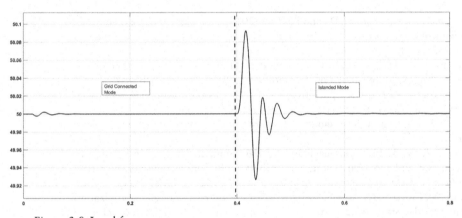

Figure 2.8 Load frequency

HRES is exposed to different load conditions to test the system. At t = 0sec, system is operating in rated load conditions; at t = 0.4s 120% of the rated load is levied on the system and at t = 0.6s load is dropped to 80% of the rated load. The response of the system can be seen as the terminal voltage is traced in Figure 2.7. From the voltage waveform, it can be observed that the level of the load voltage is preserved irrespective of any changes in source and load. The frequency of the load, as styled in Figure 2.8 is, maintained within the specified limits even when the system switches from grid connected operation to islanded mode.

Conclusion

In this article a LFC is proposed for a hybrid-renewable energy system (HRES) operating in islanded mode. The controller constants are optimized by employing particle swarm optimization (PSO) algorithm. An original search procedure is adopted to do away with the requirement for a real-time optimization method. All wind speed and irradiation values within the operational range are accommodated by the determined parameters, which are ideal. The effectiveness of the system is validated under various load scenarios, while the random nature of renewable sources is continued. The system is simulated for three levels of loads viz. rated load, above rated load and below rated load. The simulation results show a constant load voltage and frequency even after subjecting the system with variable solar irradiation, wind velocity and variable load. The competency of the controller is evident from the results obtained.

References

Alam, M. J., Hossen, T., Paul, B., and Islam, R. (2013). Modified sinusoidal voltage and frequency control of microgrid in island mode operation. *International Journal of Scientific and Engineering Research*, 4(2), 1–6.

Al-Saedi, W., Lachowicz, S. W., Habibi, D., and Bass, O. (2013). Power flow control in grid connected microgrid operation using particle swarm optimization under variable load conditions. *International Journal of Electrical Power and Energy Systems*, 49, 76–85.

Al-Saedi, W., Lachowicz, S., Habibi, D., and Bass, O. (2017). PSO algorithm for an optimal power controller in a microgrid. *IOP Conference Series: Earth and Environmental Science*, 73, 12–28.

Barik, A. K. and Das, D. C. (2020). Coordinated regulation of voltage and load frequency in demand response supported biorenewable cogeneration based isolated hybrid microgrid with quasi oppositional selfish herd optimization. *International Transactions on Electrical Energy Systems*, 30, e12176.

Bolton, W. (1997). Essential Mathematics for Engineering. Butterworth-Heinemann (Oxford).

Chandrakala, K. R. M. V. and Balamurugan, S. (2016). Simulated annealing based optimal frequency and terminal voltage control of multi source multi area system. *International Journal of Electrical Power and Energy Systems*, 78, 823–829.

Chung, I., Liu, W., Cartes, D. A., and Schoder, K. (2008). Control parameter optimization for a microgrid system using particle swarm optimization. *In IEEE International Conference on Sustainable Energy Technologies*, (pp. 837–842).

Dehghan Banadaki, A., Mohammadi, F. D., and Feliachi, A. (2017). State space modeling of inverter based microgrids considering distributed secondary voltage control. *In North American Power Symposium (NAPS)*, (pp. 1–6).

Dhanasekaran, B., Kaliannan, J., Baskaran, A., Dey, N., Tavares, J.M.R.S. (2023). Load frequency control assessment of a PSO-PID controller for a standalone multi-source power system. *Technologies*, 11(1), 22.

Duman, S., Rivera, S., Li, J., and Wu, L. (2020). Optimal power flow of power systems with controllable wind photovoltaic energy systems via differential evolutionary particle swarm optimization. *International Transactions on Electrical Energy Systems*, 30, e12270. https://doi.org/10.1002/2050-7038.12270.

Eberhart, R. and Kennedy, J. (1995). A new optimizer using particle swarm theory. *In Proceedings of the 6th International Symposium on Micro Machine and Human Science (IEEE)*, (pp. 39–43).

El-Sehiemy, R., Shaheen, A., Ginidi, A., Al-Gahtani, S.F. (2023). Proportional-integral-derivative controller based-artificial rabbits algorithm for load frequency control in multi-area power systems. *Fractal and Fractional*, 7(1), 97.

Gervasio, F. A., Bueno, E., Mastromauro, R. A., Liserre, M., and Stasi, S. (2015). Voltage control of microgrid systems based on 3lnpc inverters with LCL-filter in islanding operation. *International Conference on Renewable Energy Research and Applications (ICRERA)*, Palermo, Italy, 827–832. doi: 10.1109/ICRERA.2015.7418527.

Hoseini, S. M., Vasegh, N., and Zangeneh, A. (2020). Distributed nonlinear robust control for power flow in islanded microgrids. *International Journal of Electrical and Electronics Engineering (IJEEE)*, 16(2), 235–247.

Jamshidpour, E., Saadate, S., and Poure, P. (2015). Energy management and control of a stand-alone photovoltaic/ultra capacitor/battery microgrid. *IEEE Jordan Conference on Applied Electrical Engineering and Computing Technologies (AEECT)*, 1–6. doi: 10.1109/AEECT.2015.7360584.

Jumani, T. A., Mustafa, M. W., Md Rasid, M., Mirjat, N. H., Leghari, Z. H., and Salman Saeed, M., (2018). Optimal voltage and frequency control of an islanded microgrid using grasshopper optimization algorithm. *Energies*, 11, 3191–3211.

Karimi, H., Beheshti, M. T. H., and Ramezani, A. (2016). Decentralized voltage and frequency control in an autonomous ac microgrid using gain scheduling tuning approach. *In 24th Iranian Conference on Electrical Engineering (ICEE)*, (pp. 1597–1602).

Khooban, M., Niknam, T., Shasadeghi, M., Dragicevic, T., and Blaabjerg, F. (2018). Load frequency control in microgrids based on a stochastic noninteger controller. *IEEE Transactions on Sustainable Energy*, 9(2), 853–861. doi: 10.1109/TSTE.2017.2763607.

Kim, H.-S. and Sul, S.-K. (2011). A novel filter design for output LC filters of PWM inverters. *Journal of Power Electronics*, 11, 74–81.

Lekshmi, R. R., Hareesh, K., Swetha P., Raj, V., and Narula, A. (2016). Fuzzy gain schedule based load frequency control of multi area thermal system under deregulated environment. *In IEEE International Conference on Power Electronics, Drives and Energy Systems (PEDES), Trivandrum*, (pp. 1–6). doi: 10.1109/PEDES.2016.7914405.

Li, H., Eseye, A. T., Zhang, J., and Zheng, D. (2017). Optimal energy management for industrial microgrids with high penetration renewables. *Protection and Control of Modern Power Systems*. 2(12), 1–14.

Liu, L., Senjyu, T., Kato, T., Howlader, A. M., Mandal, P., and Lotfy, M. E. (2020). Load frequency control for renewable energy sources for isolated power system by introducing large scale PV and storage battery. *Energy Reports*, 6, (9), 1597–1603.

Mohanty, B., Panda, S., and Hota, P. K. (2014). Controller parameters tuning of differential evolution algorithm and its application to load frequency control of multi-source power system. *International Journal of Electrical Power and Energy Systems*, 54, 77–85.

Mudaliyar, S., Duggal, B., and Mishra, S. (2020). Distributed tie-line power flow control of autonomous DC microgrid clusters. *IEEE Transactions on Power Electronics*, 35(10), 11250–11266. doi: 10.1109/TPEL.2020.2980882.

Pandi, V. R., Al-Hinai, A., and Feliachi, A. (2017). Adaptive coordinated feeder flow control in distribution system with the support of distributed energy resources. *International Journal of Electrical Power and Energy Systems*, 85, 190–199.

Ramezani, M. and Li, S. (2016). Voltage and frequency control of islanded microgrid based on combined direct current vector control and droop control. *EEE Power and Energy Society General Meeting (PESGM)*, 1–5. doi: 10.1109/PESGM.2016.7741786.

Serban, I. and Marinescu, C. (2011). Aggregate load-frequency control of a wind-hydro autonomous microgrid. *Renewable Energy*, 36(12), 3345–3354. ISSN 0960-1481.

Sharaf, A. M. and El-Gammal, A. A. A. (2009). Optimal self regulating PID controller for coordinated wind-FC-diesel utilization scheme. *In Third UKSim European Symposium on Computer Modeling and Simulation*, (pp. 418–423).

Sharaf, A. M., Aljankawey, A., and Altas, I. H. (2007). A novel voltage stabilization control scheme for stand-alone wind energy conversion systems. *In International Conference on Clean Electrical Power*, (pp. 514–519).

Suchetha, C, and Ramprabhakar J. (2018). Optimization techniques for operation and control of microgrids—Review. *Journal of Green Engineering*, 8(4), 621–644. River Publishers doi: 10.13052/jge1904-4720.847

Suchetha, C. and Ramprabhakar, J. (2019). Optimal power flow controller for a hybrid renewable energy system using particle swarm optimization. *In National Power Electronics Conference (NPEC), Tiruchirappalli, India*, (2019), (pp. 1–6). doi: 10.1109/NPEC47332.2019.9034782.

Syed, I. M. and Raahemifar, K. (2014). Space vector PWM and model predictive control for voltage source inverter control. *International Journal of Industrial and Manufacturing Engineering*, 8(11), 1751–1757

Warrier, B. R., Vijayakumari, A., and Kottayil, S. K. (2020). Enhanced dynamic performance in grid tied bidirectional converter with direct power model predictive control. *International Journal of Power Electronics*, 12(3), 267–281.

Xie, S., Zeng, Y., Qian, J., Yang, F., Li, Y. (2023). CPSOGSA optimization algorithm driven cascaded 3DOF-FOPID-FOPI controller for load frequency control of DFIG-containing interconnected power system. *Energies*, 16(3), 1364.

Younesi, A., Shayeghi, H., and Siano, P. (2020). Assessing the use of reinforcement learning for integrated voltage/frequency control in AC microgrids. *Energies*, 13, 1250.

3 A thermocouple and PV-fed single inductor dual input single output hybrid DC-DC converter for standalone battery charging application

Tuhinanshu Mishra[a] and R.K. Singh

Department of Electrical Engineering, Motilal Nehru National Institute of Technology, Allahabad (Prayagraj)

Abstract

This paper proposes a renewable energy-based power conversion scheme through a dual input single output DC-DC converter. A hybrid combination of photovoltaic cells and thermocouple elements extracting heat energy from brick furnaces is proposed to establish a standalone battery charging station. The operating modes of the converter are shown. Extensive performance analysis of the configuration is done through mathematical modeling and simulation. The hardware implementation of the converter is done to justify the operation. The proposed configuration finds broad scope in the utilization of heat energy and energy conservation.

Keywords: Converters, multi-input multi-output, renewable energy

Introduction

Energy efficiency is a major contributing factor in the power sector whether in the topology design of converters or schematic of power plants. Since energy loss is equivalent to the non-utilization of existing fossil fuels, minimization of losses is essential. Some losses in the generation, transmission, conversion, and distribution are, however, implied in the process and are unavoidable.

Harvesting energy from these losses and converting it to usable forms will revitalize the energy sector. In this paper, the excessive heat energy lost in a brick furnace is converted to electrical energy with the help of Peltier devices based on thermoelectric coupling (Pham, 2020). A PV panel is also used as auxiliary support. Figure 3.1 shows the proposed schematic. This energy is supplied to a DC battery charging station through a dual input single output (DISO) converter topology (Chen et. al., 2020; Li et. al., 2020; Chen et. al., 2021; Mishra, 2022).

The DISO converter proposed in this paper derives power from a PV source and thermocouple elements (Chang et. al., 2020; Xing and Liu, 2020; Jhang et. al., 2020). It then transfers it to a battery load for charging applications. The voltage control at the load end is provided by the duty cycle control of switches. The proposed DC-DC converter topology can operate in both buck and boost modes.

[a]tuhinanshu.2020ree07@mnnit.ac.in

DOI: 10.1201/9781003450917-3

The advantages of the proposed DISO converter configuration are: the converter is based purely on waste energy harvesting and renewable energy; the proposed topology has only a single inductor implying a reduction of size and switching losses compared to several contemporary topologies; the flexibility of operation is widened since the proposed DISO converter can have either PV or thermocouples or both sources simultaneously; the converter can operate in both buck and boost modes.

Proposition of design

Proposed scheme

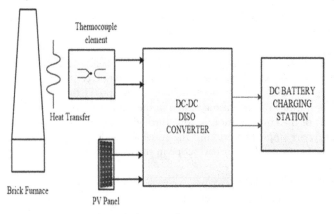

Figure 3.1 Scheme of energy harvesting

Topology of the converter

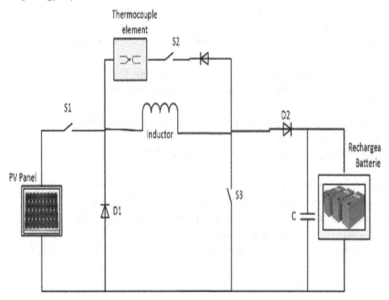

Figure 3.2 Proposed DISO converter topology

Modes of operation

The proposed DC topology operates in three modes. The converter topology and operating modes are shown in figures 3.2–3.5. In the first two modes, the inductor charging takes place by the PV source and the voltage generated by the thermocouple elements. The third mode is the discharging mode of operation of the converter. In this mode, the battery station is charged by the inductor.

Mode 1

In this mode, the PV panel charges the inductor through switches S1 and S3. The thermocouple and load are cut off from the circuit during this mode of operation.

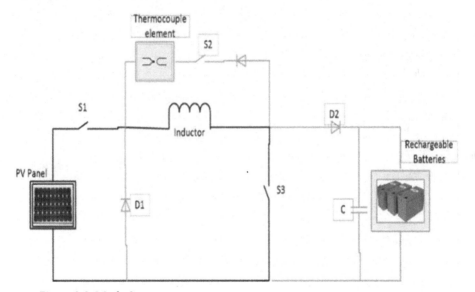

Figure 3.3 Mode 3

$$-V_{PV} + L\frac{di_{PV}}{dt} = 0$$

$$\frac{di_{PV}}{dt} = \frac{V_{PV}}{L}$$

Mode 2

The thermocouple element is activated through switch S2 and the PV source is cut off. The inductor charging takes place in this mode of operation. The inductor, already charged by the PV panel in mode 1, is now charged by the second source. Furthermore, it is to be noted that, the inductor design must be done such that it doesn't reach saturation just by single-source charging.

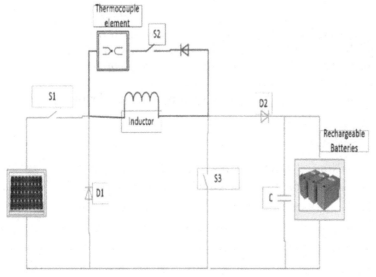

Figure 3.4 Mode 3

$$-V_{Thermo} + L\frac{di_{Thermo}}{dt} = 0$$

$$\frac{di_{Thermo}}{dt} = \frac{V_{Thermo}}{L}$$

Mode 3

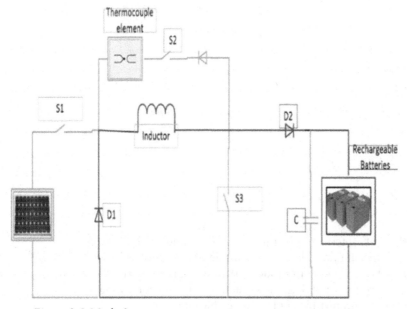

Figure 3.5 Mode 3

$$-L\frac{di_{Bat}}{dt}+E_{Bat}=0$$

$$\frac{E_{Bat}}{L}=\frac{di_{Bat}}{dt}$$

Mathematical analysis

Inductor current and voltage along with voltage gain of converter have been analysed and shown in figures 3.6–3.7.

Inductor current

Assuming, V_{Thermo} and V_{PV} are the dual sources of the converter such that $V_{Thermo} > V_{PV}$. Consider $t_1 = D_1 T_s$,

$$i_L = \frac{1}{L}\int V_L dt$$

$$=\frac{V_{Thermo}}{L}\int dt$$

$$=\frac{V_{Thermo}}{L}D_1 T_s$$

Similarly, $t_2 = D_2 T_s$,

$$\frac{V_{PV}}{L}D_2 T_s$$

For $t_3 = D_3 T_s$,

$$\frac{-V_0}{L}D_3 T_s$$

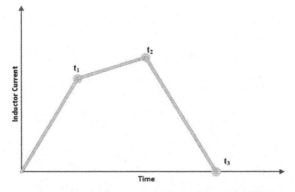

Figure 3.6 Inductor current in proposed DISO converter

Voltage gain

According to volt-second balance,

$$V_L T_{ON} - V_L T_{OFF} = 0$$

Consider the duty cycles of switches S_2 and S_3 to be a factor of the duty cycle of switch S_1.

$$D_1, D_2 = k_1 D_1, D_3 = k_2 D_1$$

where $K_1 - K_3$ are real constants.
 Assuming a relationship between the two sources,

$$V_{Thermo} = V_i$$
$$V_{PV} = x_l V_i$$

where X_1 is a real constant.
 Thus, the voltage gain is given by,

$$\boxed{\frac{V_0}{V_i} = \left(\frac{1 + x_1 k_1}{k_2} \right)}$$

Inductor voltage

The voltage levels V_1 and V_2 represent dual sources and V_3 denotes the output voltage.

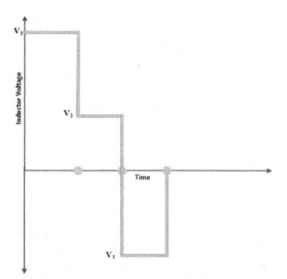

Figure 3.7 Inductor voltage in the proposed DISO

Simulation

The simulation of the proposed converter with thermocouple elements and PV source was done on MATLAB-SIMULINK and the following results were obtained. The simulation results are shown in figures 3.8–3.14.

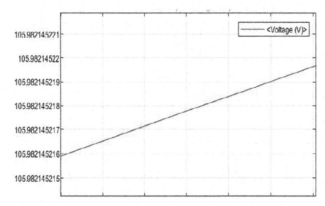

Figure 3.8 Voltage input from thermocouple source

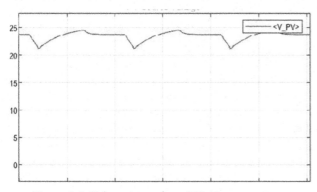

Figure 3.9 Voltage input from PV source

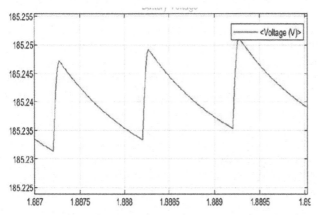

Figure 3.10 Battery voltage (charging)

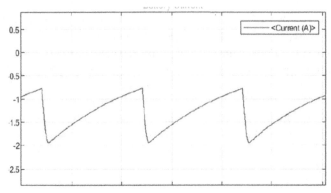

Figure 3.11 Battery current (charging)

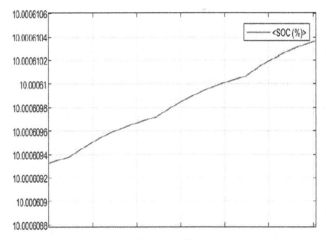

Figure 3.12 State of charge of battery

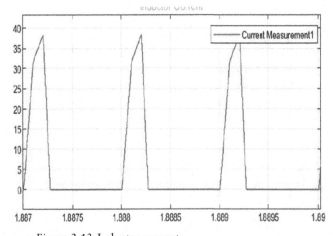

Figure 3.13 Inductor current

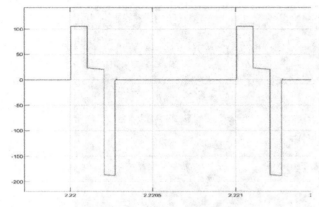

Figure 3.14 Inductor voltage

Simulation results and calculations

It is observed from the figure that the voltage output obtained ranges from approximately 185.23 to 185.25 Volts. The average inductor current is 0.639A. The ripple in output voltage is 0.01%. Table 3.1 shows the results obtained for the converter.

Table 3.1: Results obtained from the simulation

Input Voltages	V_{S1}= 105.9V	V_{S2}= 23.46 V
Output Voltage	185.24V	
Output Current	1.5A	

Hardware implementation

The average values of voltage from thermocouple and PV sources are replaced with constant DC sources of 10 volts each. The hardware setup and the obtained results are shown in figures 3.15–3.17.

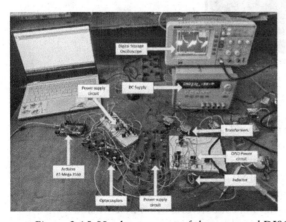

Figure 3.15 Hardware setup of the proposed DISO converter

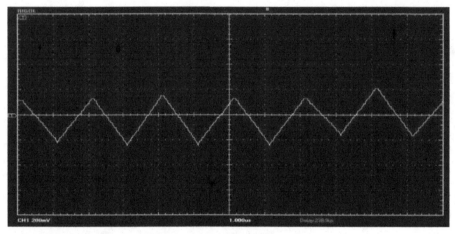

Figure 3.16 Output voltage (battery charging)

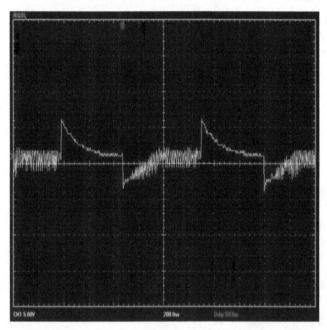

Figure 3.17 Inductor voltage waveform

Conclusion

A standalone battery charging station is proposed based on energy harvested from brick furnaces and PV panels. The sources feed a dual input single output converter topology. The proposed DISO operation is analyzed mathematically and then verified by simulation on MATLAB-SIMULINK. Finally, a laboratory prototype is built and the resulting waveforms justify the simulation.

The proposed configuration is advantageous since it converts otherwise wasted energy, using transducers, to a usable form. It is visible that the converter voltage output has a negligible ripple percentage and appreciable charging current, thus the practical implementation of the scheme will find wide scope in energy efficiency and energy conservation.

References

Chang, R. C.-H., Lei, P.-S., Huang J. K.-S., and Chen, W.-C. (2020). Batteryless DC-DC boost converter for thermoelectric energy harvesting devices. *In International SoC Design Conference (ISOCC)*, (2020), (pp. 101–102).

Chen, G., Liu, Y., Qing, X., Ma, M., and Lin, Z. (2021). Principle and topology derivation of single-inductor multi-input multi-output DC-DC converters. *IEEE Transactions on Industrial Electronics*, 68(1), 25–36.

Chen, Y., Wang, P., Elsasser, Y., and Chen, M. (2020). Multicell reconfigurable multi-input multi-output energy router architecture. *IEEE Transactions on Power Electronics*, 35(12), 13210–13224.

Jhang, J., Wu, H., Hsu, T., and Wei, C. (2020). Design of a boost DC–DC converter with 82-mV startup voltage and fully built-in startup circuits for harvesting thermoelectric energy. *IEEE Solid-State Circuits Letters*, 3, 54–57. doi: 10.1109/LSSC.2020.2978850.

Li, X. L., Tse, C. K., and Lu, D. D. (2020). Single-inductor multi-input multi-output DC-DC converter with high flexibility and simple control. *In 2020 IEEE International Symposium on Circuits and Systems (ISCAS)*.

Mishra, T. (2022). Synthesis and modelling of a novel multi-input multi-output system topology implemented on non-inverting buck-boost converter for renewable energy applications. *In First International Conference on Electrical, Electronics, Information and Communication Technologies (ICEEICT)*, (2022), (pp. 1–6).

Pham, V. K. (2020). A high-efficient power converter for thermoelectric energy harvesting. *In 5th International Conference on Green Technology and Sustainable Development (GTSD)*, (2020), (pp. 82–87).

Xing, and Liu, L. (2020). A 10mV input, 93.6% peak efficiency three-mode boost converter for thermoelectric energy harvesting. *In IEEE 15th International Conference on Solid-State & Integrated Circuit Technology (ICSICT)*, (2020), (pp. 1–3).

4 Capacity for innovation frugal: the look from companies incubated in park technology and in co-working spaces

Thais Helena Costa da Silva[1], Laodicea Amorim Weersma[1], Sandeep Kumar Gupta[2,a], Deepa Rajesh[2], Lalit Prasad[3], Priyanka Mishra[4] and Ananta Uppal[5]

[1]Centre Christ University, Brazil

[2]AMET University, Chennai, India

[3]Dr D Y Patil Institute of Management Studies, Pune, India

[4]Sri Balaji University, Pune

[5]P.P Savani University, Surat, India

Abstract

Due to competitiveness with international markets, innovation becomes imperative for the development of nations. It is a crucial resource for companies to increase competitiveness. Thus, the continual attainment of competitive advantages and at the same time reduction of the erosion that competition tends to cause, impel organizations to acquire innovative capacity. Which is how companies acquiring and save their knowledge and later turn them into innovation? Because of such arguments, this research objective is to analyses the innovative capacity for frugal innovation of companies incubated in incubators and crowded co-working spaces. In the results are presented three case studies of research environments. The questionnaire application from six companies defined the degree of the factors that lead to frugal innovation in their organizations. Finally, a radar of innovative capacity has been done, showing in which factors the companies have a greater tendency to make innovation. In conclusion, are presented the reasons why companies should invest in innovation capacity for frugal innovation and how this research contributes to academia and future research.

Keywords: Frugal innovation, frugal innovation factors, incubators, innovative capacity, radar of innovative capacity

Introduction

From the context of the internationalization of markets, innovation has become imperative for the development of nations. Thus, the continual attainment of competitive advantages and, at the same time, reduction of the erosion that competition tends to cause, impels organizations to acquire innovative capacity.

Innovative capacity is the process through which companies acquire and save knowledge and then turn them into innovation. Rocha et. al (2016) proposed that the main sources of innovative capacity are research and development sectors where new applications are produced; acquisition of knowledge through

[a]skguptabhu@gmail.com

DOI: 10.1201/9781003450917-4

the purchase of pa-tent licenses and the use of trademarks and also acquire of machinery and equipment, and training employees for improvement. However, small and medium-sized companies do not have sufficient resources to maintain a research and development sector and therefore they need to think about various ways to acquire knowledge for innovation, through informal meetings with their employees, in-formal surveys in the community and make use of social networks to get more in-formation and attract new customers. Keeping in mind the low-income market, Midgley et. al. (1978) and Hirschman (1980) developed a concept of innovative capacity and linked it with the behavior of class C and D consumers, who have limited access to education as ultra-technological items. Because of such considerations, the authors defined innate capacity for innovation as the degree to which the individual makes decisions about adopting innovation, whether in his work environment or his personal life, regardless of the influences of other people or organizations. Hirschman's (1980) concept of innovation capacity focuses on the desire of consumers to obtain information about that innovation, being considered as the inherent quest for novelty and defined as the desire to seek the new and the different. Because of such arguments, this article aims to analyze the innovative capacity for frugal innovation of companies incubated in incubators and crowded co-working spaces. The results are presented in three case studies in research environments, a questionnaire application in six companies that de-fined the degree of the factors that lead to frugal innovation in their organizations and finally a radar of innovative capacity was done, showing which factors the companies have a greater tendency to make innovation. In the conclusion, the rea-sons are presented, as to why companies should investing innovation capacity for frugal innovation and how this study contributes to academia and future research.

Innovative capacity

Understanding that innovation is a crucial tool for the success of organizations, it becomes necessary to understand how the organization can make its processes and products/services more innovative in the increasingly competitive market. Because of the above, it is pertinent to express the concepts of innovative capacity, since there is no certain definition of the term. Dosi (1998) affirms that innovative capacity is related to different degrees of accumulation of technology and different efficiencies in the process of innovative search. Go further and understand that technological capacity incorporates the re-sources needed to generate and manage technological changes and that these resources accumulate in individuals and their organizational processes (Bel-land Pavitt, 1995). Silva (2008) adds that innovative capacity is a set of features and resources that the organization produces, shares and manages that facilitate and support its innovation strategies. According to Miranda et. al. (2014), innovative capacity has a positive impact on business performance, since the number of companies that do not know how to use their resources properly or that are paralyzed in the search for knowledge, experience and technological capacity to develop products, services and/or innovative processes.

However, for firms to increase their capacity it is necessary to identify which activities influence the occurrence of such capacity, according to Weersma et. al. (2014), the following Table 4.1 was constructed based on the results of the Brazilian Technological Innovation Survey - PINTEC defines innovative activities as the company's efforts in the development and implementation of new or improved products (goods or services) and processes. In the table, the authors collected the main information contained in the PINTEC that comprises the determinants of innovative capacity. Hauser et. al. (2015) have developed a model to measure innovation capacity in technology parks based on technological, operational, inter-organizational, management, marketing and strategic resources. There was an evaluation of the external environment to analyze how parks influence and are influenced by external agents (governments, NGOs, culture etc.) and the internal environment that concerns the management of innovation by park managers and how they seek to stimulate and facilitate innovation. In the conclusions of the study, the authors point out that the main focus of the model is how the local management unit performs its activities and how it generates, identifies and develops ideas and opportunities, that is, its capacity to integrate resources, skills and abilities to innovate.

Technological parks, incubators of companies and co-working

New categories of real estate ventures emerged in Australia, Asia, Canada, and later in Europe, which was termed "business parks or business parks, terms that came from the concept of Technology Parkcre-atedinthe1950s (Collarino et. al., 2015). They defined a technological park as a physical space, infrastructure, technical knowledge, logistics, research and administrative assistance to help new companies that are created within the universities by the students to insert them into the market. In Brazil, the institution (Anprotec, 2016) defined the technological park as a complex industrial and scientific-technological base, planned, formal, concentrated and cooperative, which aggregates companies whose are based on techno-logical research developed in R & D centers linked to the park. It is an enterprise that promotes a culture of innovation, competitiveness, and increased entrepreneurial skills, based on the transfer of knowledge and technology, to increase the production of wealth in a region. According to Collarino et. al. (2015), the technology park has two main objectives: to be a seedbed to cultivate technology and function as the incubator for the transfer of academic knowledge to established companies, thus stimulating the creation of spin-offs, the development of innovative products and processes and the growth of small technology-based enterprises and serve as a stimulus to regional growth and economic development. Technology parks are created to foster the creation and growth of intensive. The technology parks, in general, have a business incubator and local research laboratories. According to Wonglimpiyarat (2014) Incubators are places where companies that are embryonic and do not have many resources to launch themselves, remain static for a certain period until they launch the market. These incubators aim to encourage the creation of new ventures, support the development of the innovative capacity of small and medium enterprises and disseminate the culture of entrepreneur-ship, focused on knowledge and

Table 4.1: Innovative capacity factors

Dimension	Factors	Degree					
		0	1	2	3	4	5
		Not applicable	Very weak	Weak	Moderate	Strong	Very strong
Cost	Substantial cost savings				1	3	2
Factor 1	Solutions that offer "good and cheap "products/services				1	3	2
Factor 2	Significant cost reduction in the operational process		1			2	3
Factor 3	Significant reduction in the final price of the product/service	1				3	2
Environment	Creating a frugal ecosystem	1			1	4	
Factor4	Environmental sustainability in the operational process	1			2	1	2
Factor 5	Partnerships With local companies in the operational process			1	2	2	1
Factor 6	Efficient and effective solutions for customers' social/environmental needs	2				2	2
Core	Concentration on the main features and performance of Products / services				1	1	4
Factor 7	Focus on core product/service Functionality instead of additional functionality	1			1	3	1
Factor 8	Ease of use of the product/service			1		1	4
Factor 9	The question of the durability of the product /service (does not spoil easily)				1	1	4

Source: Research Data (2018).

innovation. According to (Storopoli et. al., 2015) for a company to be inserted inside an incubator, it is necessary to go through the incubation process that takes place in five stages: formulation of the idea, what will be the company and which product will be marketed; recognition of opportunities, where the environments of the future company and the macro-economic variables will be analyzed. Planning how to put the company into practice in the market, usually with the construction of the business cloth; entry where the company presents its business plan to the incubator that will analyze it to verify if the business is feasible and finally approved the business plan, the company installs itself in its new place where you can enjoy the infrastructure and available resources of the incubator to boost your company.

Anprotec (2016) aware of this relationship, proposed a new model called Quadruple Helix, where one more agent is inserted beside the three already known: society (people)who through their ideas, feedback and even monetary contributions to financing platforms contribute to the development and dissemination of innovations within these companies.

About Brazil, (Anprotec, 2016) shows that the incubators are classified according to the types of enterprises that shelter: 1) technology-based incubators; 2) traditional incubators; 3) mixed incubators; 4) social incubators; and 5) cooperative incubators. Technology-based incubators host companies whose products, processes or services result from scientific research, in which technology represents high value and aggregate; According to the Anprotec Technical Report (2012), completed in 2011, 40% of the incubators in the country are technology-based, 18% are traditional, 18% are mixed, 8% of services, 7% are social, 7% are agro-industrial and 2% are cultural. a case study was carried out in three environments.

Methodology

The research method was developed to carry out the research and the achievement of the established objectives. To analyze the capacity for frugal innovation in the company's understudy. It is necessary to present the environments in which these companies have inserted briefly. The first part of the research is Technological Development Park (PADETEC), the co-working space of a Fortaleza Development Bank and the INTECE incubator.

In the second part of the research, a questionnaire had made to collect the following information about the characterization of the respondents, characterization of the company, typology of innovation and what factors of innovation were present in that company. In this last questionnaire item, the factors and dimensions of frugal innovation have presented to the respondents in the five-point Likert scale format, where they will assign scores from 0 to 5 according to what they perceive to occur in the company, the data collection instrument has elements that characterize this study as mixed approach research, which uses two approaches: qualitative and quantitative. According to the guidelines of Sampieri, et. al. (2013), the quantitative approach uses statistical measurement and statistical analysis. In this case, it is referring to the Likert scale and the frugal innovation radar. The qualitative approach uses da-ta collection without statistical measurement to discover or improve research questions in the interpretation process, such as the characterization of the respondents and the company.

The research environment of this monographic study consists of the Technological Development Park of the Federal University of Ceara, PADETEC, in the city of Fortaleza, a co-working office located within a development bank and the incubator Intece. PADETEC had inaugurated in 1991 to be a business incubator and an R & D center for the generation of technology-based companies (Figure 4.1). The co-working office analyzed in this research is located in the city of Fortaleza within a development bank that aims to encourage innovative entrepreneurship and facilitate the management of innovation in the Northeast region. The selection process takes place through the registration and analysis of the business plan, the rules are that startups must have CNPJ, a consolidated team, have some innovation in their products or the production process, and their products/services must have some social impact in the areas of public interest, impact on the regional development of the Northeast and potential integration with other startups (Figure 4.2).

To participate in the program; the entrepreneur /entrepreneur must prepare the company's Business Plan proposal, presenting it to the incubator's coordination, at the CENTEC Institute's SEDE or to one of the managers of the INTECE Units located in the interior of the state.

The study is mixed research that uses the two research approaches: qualitative and quantitative. It has added that the qualitative aspect concerns the identification of possible factors that lead to frugal innovation in companies from Ceará that have in-stalled co-working spaces and incubators. Regarding the quantitative nature, we have a measurable study of how much each factor contributes to this type of innovation, in contrast, constructing the radar of factors that contribute to the frugal innovation can occur. The sample used in this work was: 1. Company of PADETEC in a universe of 13 companies that are currently incubated, 2 companies from a universe of 10 companies inserted in the space of co-working of a bank of strength development and, 3. companies that are participating in the process. The instrument of data collection will be through a questionnaire that is applied in an interview with the managers of the companies, this questionnaire uses questions about the characterization of the respondents, and the characterization of the company and in the last part, there is a Likert scale of five points.

Results

This section includes the presentation and discussion of the results of multiple case studies carried out in the Technological Development Park, PADETEC, and the co-working eco-system of Ceará from October to November 2018. The results include the application of three interviews with questionnaire support. The profiles of the respondents indicate reasonable knowledge about the subject, since many are aware of the concept of innovation, but are not aware of the term frugal innovation, which, as the theoretical framework shows, is a new definition that emerged in the 1990s. The first part of the questionnaire is the characterization of the respondents that seeks to identify the profile of the person who

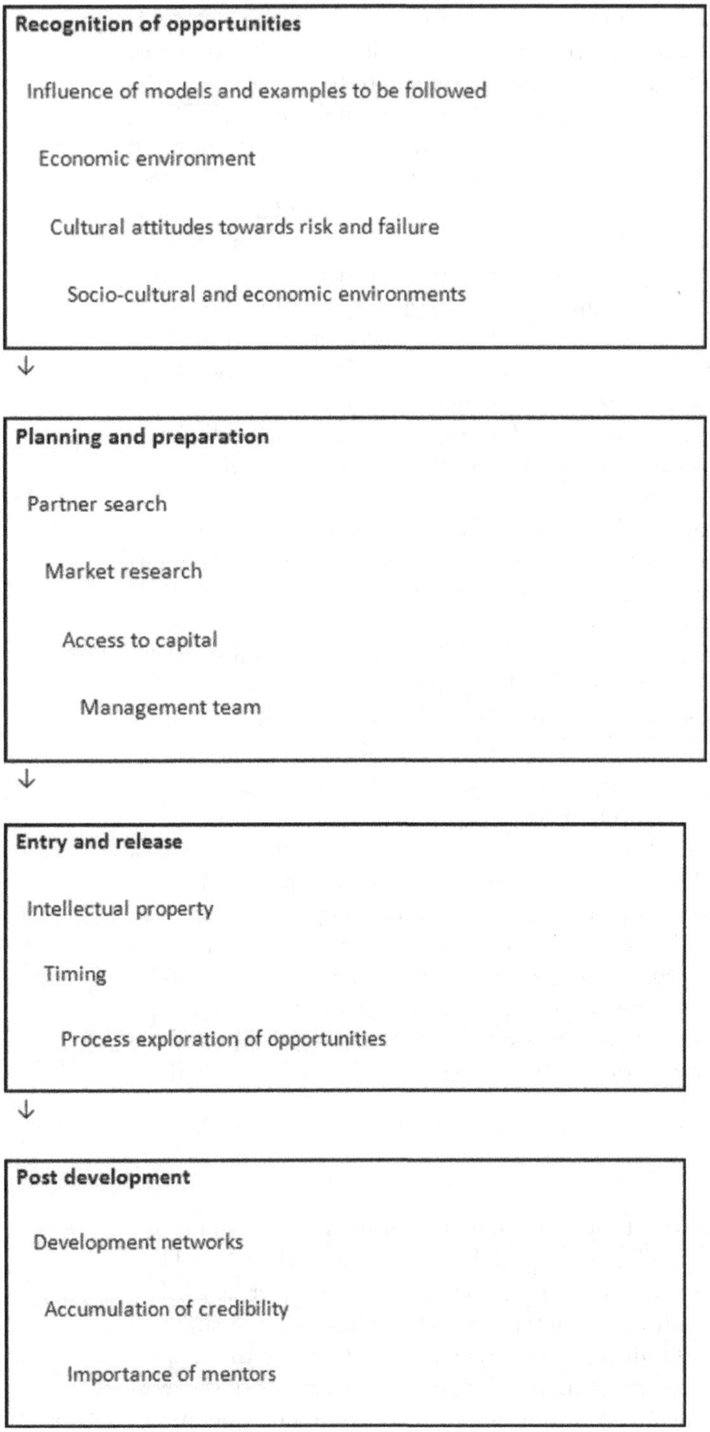

Recognition of opportunities

Influence of models and examples to be followed

Economic environment

Cultural attitudes towards risk and failure

Socio-cultural and economic environments

Planning and preparation

Partner search

Market research

Access to capital

Management team

Entry and release

Intellectual property

Timing

Process exploration of opportunities

Post development

Development networks

Accumulation of credibility

Importance of mentors

Figure 4.1 Incubation process
Source: Adapted from Hannon (2003)

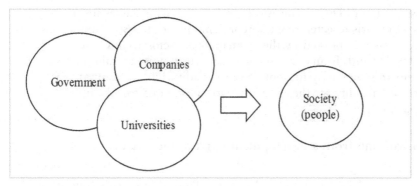

Figure 4.2 Quadruple propeller
Source: Anprotec (2016).

answered the survey, of the six interviews 4 of the respondents said that they are part of the top management of the company: the first identified as a partner and one of the owners of the business and is currently operations manager, the second and third are partners and founders of the companies and act as executive directors and the fourth is a founding partner and acts as a commercial manager. In the two remaining companies, the interviewees were students who actively participated in the production of their projects and answered that their companies were born in public schools of technical education in the interior of Ceará developed in a partnership between students and teachers. Regarding schooling, the former has a full superior and is currently doing postgraduate studies, the second is attending higher education, the third and fourth have a full superior and the fifth and sixth are still finishing high-school. Regarding the time of performance in the company, all respondents are 2 years in their companies.

Characterization of analytical units: technology development park-PADETEC

The manager of the interviewed company that is incubated in PADETEC said that he decided to seek support in the incubator because of the advantages that this type of space offers, but the main reason was to reduce costs because there the company pays a small fee as if it were a rent per shed and thus gains a physical ad-dress that offers a meeting room, research center with 1500 m², 6 research and development laboratories, 2 analytical centers with 6 laboratories, 3 pilot plants and 1 bakery unit.

From a new look, the company visualized an APP where the traveler could have the world in the palm of their hand. It was in this context of needs and with the world in mind that Wander pathswasfoundedwiththeaimofprovidingamoder-nanddynamicplatformcomposed of responsive and application site to serve customers and partners, considering the rent of the cellphone as the main Internet access device.

The shortage of resources associated with the need for better company disclosure and increased downloads in the App made the company use creativity and launch the First Contest of Digital Tourism Tour of Ceará, whose assessment bank was composed of tourism secretaries and representatives of the State tourism segment. The contest was praised by the participants, disclosed on Globo.com having generated demands from other municipalities, including capital from another state. The awards ceremony took place in the auditorium of SEBRAE, which provided greater visibility for the company with the conquest of the first customers of the plat-form

Eco-system of co-working from a development bank in the city of Fortaleza

The two companies located in the space said that the biggest reason to seek support in a co-working office was the great networking that this type of space has to offer with other companies, besides the incentive of the bank where it is located and the connections made possible by this environment.

The first company is a Chatbot-maker, focused on high technology. Its main objective is to offer other companies that use social networks as a source of sales and customer relationships, the chance to leverage their sales and capture new customers through chat-bots which are computer programs that automate converzations within the messaging applications simulating a human. Chat-bots function as virtual artificial intelligence robots that talk to customers and simulate the behavior of humans; they are programmed to understand questions, provide answers and perform tasks. Given these advantages, the customer saves time and money since the botsare is available 24 hours a day and can serve customers at any time in a standardized way while capturing information from that customer to offer exactly what he seeks. Generally, companies that use Facebook and Instagram a lot as communication networks, hire this service to optimize their demands.

The second company is Total cross a development platform designed to help customers develop good mobile applications with Java Language, using the full benefits of portability between devices such as iPhone, Android, Windows Phone 8, Windows Desktop and Linux. The advantages of this platform are real portability that makes the application have the same behavior, usability and user interface across all deployed platforms, even on desktops. Its managers sought a space of co-working because of the connection and learning opportunity with the other companies, manager realizes that in his service there is innovation, and considers his degree of innovation relevant and his type of innovation as incremental.

Incubator of the institute technological teaching center-INTECE

Secretariat of Science, Technology and Higher Education of the state of Ceará (SECITEC), in partnership with the state government, is promoting the second edition of the knowledge fair in 2018, from 21 to 24 November, intending to promote a single space, integrated, able to show the full capacity of an entrepreneurial, creative and innovative people. The first is to Remodel, reform is possible a business of social impact in housing, which offers qualified solutions in architecture to solve some of the problems of the average and low-income public who wish to renovate their homes, but do not have the capital for a large project bearing. The motivation of the managers to find a space of co-working was the co-existence with other groups of businesses, alida the reduction of costs.

The second is Bioquitina, a company that produces a powder called chitin that is extracted from shrimp cephalothorax and used to remove or absorb oil from water, a very innovative and useful product that in the future will contribute to the reduction of impacts caused by environmental disasters. Researchers sought a co-working space to reduce costs and find investors or partnerships with shrimp-processing companies. The third is Azadiract in a natural insecticide and bactericide extracted from the Azadiracht in a plant that possesses properties capable of killing human harmful insects added to cleansers that enable customers to use this product at home and in clothing as a kind of natural repellent. The researchers opted for co-working for the need for mentoring in some is soft. The company and the pursuit of investments and accelerators.

Frugal innovation capacity

Regarding the capacity for frugal innovation, the model of (Rossetto et. al., 2017) in which the respondents identified the degree of each factor according to the perceived relevance.

As shown in the table above, the responses of the three respondents to Dimension 1 that include factors one, two and three the one that obtained the highest degree of relevance, talk about cost reduction. According to Zeschky et. al.(2014) one of the main characteristics of low-cost innovation is the reduction of costs.

Dimension 2 considers factors four, five and six as the ones that obtained less degree of relevance or that had greater dispersion of points, it talks about trans-forming the external environment where the company is inserted in a frugal ecosystem, through actions in the process of operational partnerships that contribute to the preservation of the environment, partnerships with local companies to share knowledge or increase sales and capture new customers and solutions that the product/service offers to meet the needs of customers who care about the environment.

Dimension 3 contemplates factors seven, eight and nine that speak of the focus on the main purpose of the product or service offered, because at first a product can be created for a certain purpose, but if it proves very effective for another function. One of the characteristics that make frugal innovation accessible to all audiences besides reducing costs is the possibility of using simple raw materials and transforming them into products.

Radar of innovative capacity

To compile the data obtained from the factors belonging to the dimensions of substantial cost reduction, creation of a frugal ecosystem and focus on the main functionalities of the products/services, we sought to build the innovative capacity radar as shown in the following chart.

Radar aims to simplify decision-making and make it more assertive to managers, this model of chart allows a simplified view of the degree of importance attributed to each factor by the respondents. The radar was constructed based on the responses given in the framework on frugal innovation capacity, respondents were assigned scores of 0 (not applicable) represented by pink to 5 (very strong) color represented by green color, dimensions and their respective factors included in the table.

For a good understanding of the graph, it is important to pay attention to the lines where numbers 1, 2 and 4 appear because the questionnaire has six re-pendants. The factors that appear with colored lines in the last line with the number 2, mean that two of the companies interviewed marked as the core from 0 to 5 where each one is represented by a color in that factor, the same thing happens with the lines that appear with the number 1 and 4 means that one or four companies inter-viewed gave a score of 0 to 5 for each factor. These results corroborate with the theoretical basis of (Miranda et. al., 2014), which explains the difficulty that companies have to identify which elements contribute to making them more innovative.

The manager of the company Levarti was able to assign notes and respond to all dimensions with clarity, in size 1 the manager said that his service to companies is good and cheap and there is a reduction of costs in the operational process, but in the final product, he has no control of the price, since the individual client closes the package of travel with partner companies. In dimension 2 he said that the company contributes to the creation of a frugal ecosystem, as the company helps in raising awareness of its partners and customers about the

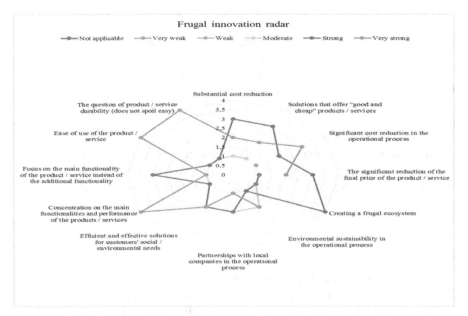

Graph 4.1 Frugal innovation radar
Source: Research data (2018).

environment and in dimension 3 the service comes with several others, so the focus goes beyond the main function, your services do not spoil and your application and your site are easy to use.

The direct executive of remodeling assigned scores of 4 and 5 to all dimensions and their respective factors, as their innovation is more explicit in the process and in the way their service is rendered to the average and low-income public, the company together with its partners who offer the construction material needed for the renovation of houses can reduce costs in the operational process and thereby reducing the final price of the service. In dimension 2, which refers to the ecosystem, the company assigned its only 3 (moderate) rating on the environmental sustainability factor in the operational process, since in the execution of the reform are the partners or outsourced people who do the work, and remodeling does not verify if the products used by them are sustainable or not. Already the factors concerning partnerships with local companies and attention to the client's environmental and social needs to note were strong and very strong.

In dimension 3 which relates to the main functionality of the products, the notes were strong and very strong, as the focus is on the main service which is the reform of the house, the use of its service is easily accessible, by the company web-site where the customer makes the registration and the service does not spoil. The-company Bioquitina attributed a very strong degree to most of the dimensions and their respective factors, since its product is still being developed in a handmade way does not harm the environment and reduces the costs both in the manufacturing process and in the final price of the product that does not spoil and it is easy to use, the chitin powder has other functionalities besides being absorbing oil from the water, but this is its main functionality, however in matters of partnerships with companies to help in the operational process the Bioquittin attributed weak grade, the reason that made her look for a co-working space. The company Azadiractina also attributed a strong and very strong degree to most dimensions, its product is produced in a handmade way through the extraction of a plant of the same name, therefore it does not damage the environment because it does not have chemical compounds in its formula and consequently reduces the price of the final product, it is easy to use and can be used in cleaning the house or in clothing to prevent mosquitoes and kept many times because its durability is high. Frugal innovation capacity and impact on innovations developed by respondents, From the data found through the interviewees' answers, it was noticed that the innovation capacity of the companies received a higher score in factors one, two and three which refers to cost reduction since all six companies stated that there was a reduction in price end of the product or its production process. Factors four, five and six were given the lowest score because companies offer technology services through mobile applications or internet sites, which does not have as much impact on the environment. The factors seven, eight and nine received medium scores because as it deals with services, the durability does not have an expiration date, since they are applications and platforms of the Internet, the clients already have a certain facility for handling them and the focus in the functionalities is greater than in the additional features of the products. What these two companies offer impacts the lives of their customers and change the way they communicate; this change is the essence of the innovation offered by them.

In addition to putting everything in one place, the way to innovate is also on the humanitarian side of the company with its main service is to create marketing campaigns for local small businesses in Brazil, mostly from Ceará that does not have the money to advertise their services and offer these clients with the support of the state government of Ceará, training on how to treat clients and the impact of their actions on the environment, as Ceará as a coastal region must preserve its beaches and one of its greatest sources of income: tourism. The product proposed by the company Bioquitina is inserted in the context that Freeman and Soete (2008) describe as innovations capable of causing a change in the economic markets and consequently trans-forming the life of a nation. The simple fact that the company uses a "raw material" that is the head of shrimp, which for most people is seen as disposable already makes its innovation as being frugal and its usefulness is of great importance for any country, but mainly for Brazil, where lately we see environmental disasters happening more frequently, an example is the oil spill in the municipality of Mariana in Minas Gerais. The product of Bioquitina can be used to mitigate the impacts of tragedies such as Mariana and prevent the pollution from becoming bigger where in the future the concept and principles of sustainability no longer exist.

The product developed through the Azadirachtin plant to become a natural insecticide is frugal innovation in the way it is artistically produced, Tran and Ravaud (2016) defined four types of frugal innovation that describe the most sim fewer bottom innovations that are the innovations that use little or no technology, such as Azadirachtin. This product, as simple as its functionality may seem, can have a huge impact in areas where the environment is in the most precarious situations. The low-income public would be the most benefited by this product, the insecticide does not have the power to alone eliminate all harmful insects, without a proper hygiene stem there will always be many insects, especially those that attract by dirt but an insecticide is sold at an affordable price, is already a great help to help in control.

Conclusion

In the conclusion of this study, it had noted that the environments presented as technological parks, co-working and incubators have a greater tendency to promote the innovative capacity in their companies. Since one of the objectives of these spaces is to promote innovation. In the description of the environments, it is possible to find in their organizational pillar practices that encourage companies to innovate to do constant research, invest in human capital, invest in new equipment and others. Public and private companies that act in the trade hardly promote such actions because many of them believe that they do not need innovation or it will have a very high cost. It has already found that companies risk more in the field of innovation. The case studies, it has pointed out that one of the fundamental rules for companies is to be in an incubator and a technology park. It is to have innovation in some part of their company and to innovate the product, or in the organizational process, the co-working spaces encourage innovation which relates to the sharing of knowledge with other companies that occupy the same space. The dimensions and factors of frugal innovation presented (Rossetto, 2017) address the reduction of costs, the environment, where innovation has been inserted, and how frugality influences the functionalities of the product or service. Which have the primary functions of the product or service can be used by any user regardless of class, race, or purchasing power, among other characteristics.

In the companies studied, four existing services that have already been offered in the market, but in a different way like Chat-bot maker, Total cross, Levarti and Remodeler re-duce the costs. So, the consumer can have access to something of quality without paying extra for it. Remodeler and Levarti have partnerships with other smaller companies, which help build a frugal environment and promote care for the environment to help the community, where they are inserted, through donations and actions that show the importance of the com-munity to the customers. The remaining two companies Bioquitina and Azadiractin-an offer products that do not harm the environment and that have a hand-made production. Natural products make their products. This influences the final price of the product which is quite affordable and these products have been developed in public schools in the interior of Ceará. Where they did not have many resources available for research. This research is essential for the academic world in which cafe papers talk about frugal innovation indicator and their importance. Knowing, how to measure frugal innovation will encourage companies to invest more in practices to reduce costs; will raise awareness of the environment for the community is an essential part to make a business thrive. Finally, as a suggestion for future studies, during this research, it was noticed that several relaxed citing innovations have developed in environments that do not receive as many investments from companies or the government like Bioquitina and Azadiractin. It has developed in public schools, but they need the capital to grow as startups. In these places even with few teachers have tried to encourage students to innovate, despite the lack of resources and often do not receive the attention, that they have deserved.

References

Helio Gomes da Rocha Neto; Robert A Rosenheck; Elina A Stefanovics; Maria Tavares Cavalcanti (2016). Attitudes of Brazilian Medical Students Towards Psychiatric Patients and Mental Illness: A Quantitative Study Before and After Completing the Psychiatric Clerkship,

da Rocha Neto HG, Rosenheck RA, Stefanovics EA, Cavalcanti MT. (2016). Attitudes of Brazilian Medical Students Towards Psychiatric Patients and Mental Illness: A Quantitative Study Before and After Completing the Psychiatric Clerkship. Acad Psychiatry. 2017 Jun;41(3):315-319. doi: 10.1007/s40596-016-0510-6. Epub 2016 Feb 16. PMID: 26883528.

David F. Midgley, Grahame R. Dowling (1978), Innovativeness: The Concept and Its Measurement, Journal of Consumer Research, Volume 4, Issue 4, March 1978, Pages 229–242, https://doi.org/10.1086/208701

Hirschman, E. C. (1980). Innovativeness, novelty seeking, and consumer creativity. Journal of Consumer Research, 7(3), 283–295. https://doi.org/10.1086/208816

Dosi, G. (1998). Sources, Procedures and Microeconomic Effects of Innovation. Journal of Economic Literature, 26, 1120-1171.

Bell, M., & Pavitt, K. (1992). Accumulating Technological Capability in Developing Countries. The World Bank Economic Review, 6, 257-281.

Silva, Maria José, João Leitão and Mário Raposo (2008). Barriers to innovation faced by manufacturing firms in Portugal: how to overcome it for fostering business excellence?, Int. J. Business Excellence, Vol. 1, Nos. 1/2, pp. 92-105.

Weersma Amorim Laodicéia, Jorge António Barbosa Ferreira, Arnaldo Coelho (2014), The mediating effect of strategic orientation, innovation capabilities and managerial capabilities among exploration and exploitation, competitive advantage and firm's performance, Accounting & Administration, 64(1), http://dx.doi.org/10.22201/fca.24488410e.2019.1918

Collarino A, Vidal-Sicart S, Perotti G, Valdés Olmos RA (2015). The sentinel node approach in gynaecological malignancies. Clin Transl Imaging. 2016;4(5):411-420. doi: 10.1007/s40336-016-0187-6. Epub, Jun 13. PMID: 27738629; PMCID: PMC5037154.

Anprotec (2016). ANPROTEC – Associação Nacional de Entidades Promotoras de Empreendimentos Inovadores. Disponível em:<http://anprotec.org.br/site/>. Acesso em: 02 maio.

Wonglimpiyarat, J. (2014). Competition and Challenges of Mobile Banking: A Systematic Review of Major Bank Models in the Thai Banking Industry. Journal of High Technology Management Research, 25, 123-131. https://doi.org/10.1016/j.hitech.2014.07.009

Hannon Lance (2003). Poverty, Delinquency, and Educational Attainment: Cumulative Disadvantage or Disadvantage Saturation?, Sociological Inquiry, 73(4), pp. 574-595, https://doi.org/10.1111/1475-682X.00072

Anprotec Technical Report (2012), https://www.certi.org.br/en/files/RA-CERTI- 2012-EN.pdf

Sampieri, R. Hernandez C. Fernandez Collado, P. Baptista (2013). Research methodology, https://www.esup.edu.pe/wp-content/uploads/2020/12/2.%20Hernandez,%20Fernandez%20y%20Baptista-Metodolog%C3%ADa%20Investigacion%20Cientifica%206ta%20ed.pdf

Zeschky Marco B., Stephan Winterhalter, and Oliver Gassmann (2014). From Cost to Frugal And Reverse Innovation: Mapping the Field and Implications for Global Competitiveness, Research-Technology Management, 57(4), DOI:10.5437/0856308X5704235

Freeman and Soete, Freeman, C. and L. Soete (1997). The Economics of Industrial Innovation London: Francis Pinter.

Tran and Ravaud, T. Tran, P. Ravaud (2016). Frugal innovation in medicine for low resource settings, BMC Medicine, 14:102, DOI 10.1186/s12916-016-0651-1

5 A fuzzy MCDM approach to create a road map for Industry-4.0

Peeyush Vats[1,a], Surya Prakash[2] and Rekha Nair[3]

[1]Department of Mechanical Engineering, Poornima College of Engineering, Jaipur, Rajasthan, India

[2]Department of Supply Chain and Operations Management, IIHMR, Jaipur, Rajasthan, India

[3]Department of Applied Science and Humanities, Poornima College of Engineering, Jaipur, Rajasthan, India

Abstract

Industry 4.0 (I4.0), a forward-thinking manufacturing concept, seeks to enhance the speed and adaptability of industrial systems while revolutionizing the manufacturing industry for the future. This cutting-edge approach incorporates advanced technologies like cloud computing (CC), big data concepts (BDC), wireless systems (WS), the Internet-of-Things (IoT) and cyber-physical-systems (CPS) into the manufacturing process. I4.0 is viewed as a multidimensional concept as it involves various strategic decisions that affect multiple attributes during the decision-making process. The main goal of this study is to prioritize the selection of suitable strategies for I4.0 using a multi-attribute decision-making (MCDM) approach in an. uncertain and ambiguous environment using the inputs of multiple experts who provide their opinions on critical decision aspects. This study proposes a fuzzy multi-attribute decision.-making (FMCDM) approach that utilizes similarity to an ideal solution to address the problem and enhance the applicability of uncertain data.

Keywords: Fuzzy TOPSIS, Industry 4.0., MCDM approaches

Introduction

It has been assumed that the earliest age of industrialization originated in the mid eighteenth-century and there will be a conversion of society from the agricultural world to the industrial world. Automation systems have been incorporated into the production system to address the uncertain demands of the growing population. The involvement of machines has been included to produce steam from water at the end of the eighteenth century (Kagermann et. al., 2013). In various articles, it was treated as the first industrial revolution (Qin et. al., 2016). The utilization of assembly lines have been started in 1870 in the industry and this period was designated as industry 2.0 (Schla pfer et. al., 2015). Womack et. al. (1990) stated that, at the starting of the 20th century, assembly lines became more popular, and Ford applied them to its companies. Due to the worldwide spread of digitalization and the application of PLC, industry 2.0 has been replaced by industry 3.0 (Segovia and Theorin, 2012). The adaption of computers has been involved drastically to control the production process

[a]peeyush.vats@poornima.org

DOI: 10.1201/9781003450917-5

(Lasi et. al., 2014). Since then, investment and opportunities for industrial and industrial commodities have increased simultaneously. Industry-4.0 (I4.0) may be designated as the 4th age of industrialization is a comparatively new concept in the production systems which can be revealed from the integration of the newer technologies. The main constituents of I4.0 are cloud.-computing (CC), the Internet-of-Things. (IoT), cyber-security (CS), big-data-analysis (BDA), smart robots (SR), augmented reality (AR), data exchanging (DE), simulation, etc. There is the contribution of politicians in the evaluation of the selection of suitable strategy phase from industry 3.0 to I4.0 from the starting of the 2000s along with the scientist. The conclusion is to raise the world through the deterioration of the overall industrial system and the transformation of technology into the results of I4.0 (Yang et. al., 2017). Therefore, for this evolution, it is recommended that several strategies must be adopted to select and scientifically put in for the selection of suitable strategy. In this sequence, the main objective is to choose the most suitable strategies implemented for I4.0, which possesses many attributes that affect quality and quantity, as well as many strategies that need to be identified, analyzed, and evaluated.

various strategies have been proposed and adopted to apply the. fundamental concept of I4.0. Here, the main attraction is that the traditional manufacturing business model does not meet the requirements of the emerging I4.0 technology. Here the. important points to note are that the safety, reliability, stability, and connectivity between the machines, to maintain the consolidation of various manufacturing processes, blocking the explosions of IT, to protect industrial technical knowledge are the important factors. With these issues in mind, we have adopted strategies that are flexible enough to respond to change (Sung, 2017). Fuzzy multi-attribute decision making methods (FMADMM) are recommended to select the most appropriate strategy in business cases (Prakash et. al., 2021).

Problems in the real world are often complex and there can be no single attribute to make the best decision. There are five key elements to multi-attribute decision making methods (MADMM). MADMM is typically a group of choices, options or alternatives, a collection of attributes, the weights of those attributes, ratings of various choices, and a decision matrix. There are various attributes (criteria's) and options (alternatives) to create a roadmap for I4.0. Due to these attributes (criterion's) and options (alternatives), the implementation of I4.0. becomes a multi-attribute decision-making MADM problem. In this regard, the number of attributes (criterion's) in the selection process will be investigated to establish a possible set of strategies, gather appropriate information on the strategies for the attribute, and assess their objectives using the MADM methodology (Tzeng and Huang, 2011). Incidentally, the proposed MADM technique has been combined with intuition fuzzy set theory (FST) to overcome the uncertainty created by linguistic expression. In this study a total of 11 attributes (criterion) and 14 options (strategies) have been considered to prioritize the options by applying fuzzy TOPSIS. This study has been organized as: Section 1 describes an introduction to this study. Section 2 represents the relevant literature review of different perspectives and applications of Industry 4.0). Section 3 provides the detailed description of proposed methodology of FMADMM. Application of proposed methodology and the discussion of results are described in Section 4. Conclusion is presented in section 5.

Brief review of literature

One of the important subjects for research in the current scenario is I4.0 and it became so much popular and a trending field for the people related to academic and research. In current times, the. numerous research papers have been published in the relevant field and the number of publications has increased dramatically in recent years. In literature, many researchers have provided great importance to the identification and prioritization of strategic industries. For example, Schumacher et. al. (2016) created a model that contains nine frontiers. and sixty-two attributes (criterion). to assess the growth of I4.0 with a focus on leading technologies of I4.0 for various organizational dimensions. In this model, a redesign strategy is used using nine dimensions, including commodity, operations, technology, administration, culture, employees, and policy. Chen and Huang (2017) described I4.0 and challenges, referring to redesigned, improved informational systems and organizational strategies.

Moktadir et. al. (2021) assessed the shortcomings of the application. of I4.0 in Bangladesh's leather industry. Their findings show that a lack of technological infrastructure is one of the most important hindrances for deployment of I4.0. Luthra and Mangla (2018) discovered 18 challenges and evaluated them by the Analytical Hierarchy Procedure approach for I4.0. In I4.0, market conditions, numerous technology and production process solutions, and the choice of them are investment challenges, especially given the differences in strengths they offer their consumers. Several studies review the selection of suitable strategy process to the company's qualifications and technical requirements (Hermann et. al., 2016; Narula et. al., 2021).

One more example is Priya and Malhotra (2019) in which problem-solving using MCDM approach, for strategies appropriate to I4.0 like intelligent web interface and 5G communications. Identify appropriate digital technologies in manufacturing, consider I4.0 concept, I4.0 comparison project implementation. of legacy and new systems and select the strategy best suited for implementing I4.0 (Erdogan et. al., 2018; Ramos et. al., 2020). Through research, it has been seen that most of these applications are used in limited circumstances or as part of a production. environment, such as identifying appropriate. stakeholders. The analysis of the findings of the related study of the deployment of I4.0 show that the I4.0 is the integration of different requirements of strategical and operational areas, including technology such as big data analytics, intelligent machines, sensors, IoT, vertical and horizontal system integration.

The purpose of the deployment of the concept of I4.0 was investigated from various industrial sectors. The reasons for improving the production line control of large-scale customization modes, automation systems, can be used (Simon et. al., 2018). The dimensions of the selection of suitable strategy are presented from multiple perspectives through the question and its source (Schumacher et. al., 2016). Most risks are focused on security risks that an attacker can take note of, gain unauthorized access to and change the system, issue improper commands, and cause damage or system damage (Condry and Nelson, 2016). It is concluded that the current IoT protocol does not provide real-time behavior of system functions for inter-device communication. One example of risk treatment is to capture the uncertainty and vagueness in a model and use the FST tools in MADM. The fuzzy logic has been implemented in various areas of uncertain environments in current applications.

It has been implied from previous studies that some strategies have been classified publicly from the perspective of a given dimension not generic risk and uncertainty treatment. Therefore, it cannot be used for projects of special purposes. Therefore, by referring to a variety of resources, one can attempt to determine the strategy that can fill the research gap. The FST is one of the strongest tools to describe the judicial and subjective judgment of decision providers under the conditions of fuzziness and uncertainty. Due to the increasing applications of fuzzy technology universality under uncertainty is widely implemented for the selection of suitable strategy of strategies in I4.0 in the given study.

At the end of the process of reviewing the literature, the preeminent highlights of this article may be pointed out as:

The Fuzzy TOPSIS method has been implemented to provide the ranking to the strategy of I4.0.

It also provides the direction for those organizations who want to select the most suitable strategy applied in I4.0.

This article also provides the insights to researchers and managers who are willing for finding of suitable strategies of I4.0.

Proposed methodology

For solving. the problem. of multi-attribute decision-making, Chen (2000) proposed a method called "Fuzzy TOPSIS". The Fuzzy-TOPSIS approach is has been widely used in various multi-attribute decision-making studies because it is consistently implemented in prioritizing and evaluating the different strategies of I4.0. Recently, Chen and Huang (2017), Medic et. al. (2019), Simon et. al. (2018), Kaya et. al. (2020) have been widely accepted the fuzzy methodologies to assess the different types of attributes and strategies in I4.0. Therefore, the Fuzzy TOPSIS approach is implemented to prioritize the impact of various attributes on different strategies.

The first step is to assign weights to different attributes. These scores were finalized after rigorous brainstorming sessions by experts and authors in an assimilation manner. Prejudice in judgments has also been identified and eliminated. The problem of bias is solved by analyzing all the members of the study separately in two parts. Finally, these inputs from the two sub-teams are grouped in a group. Eliminate any ambiguity and disagreement by consulting with the most experienced team members. These responses have been translated into linguistic terms (fuzzy triangular numbers) to determine the impact of different approaches on various options. To execute this study the complete procedure has been adopted from Junior et. al. (2014). The weights of attribute in linguistic scale have been classified in to five categories such as, Very_low [V_L] (0.000, 0.000, 0.250),

Low [L] (0.000, 0.250, 0.500), Exactly_equal [E_E] (0.250. 0.500., 0.750), High [H]. (0.500, 0.750, 1.000) and Very_high [V_H] (0.750, 1.000., 1.000). The linguistic scale details for evaluating attribute weights are shown in Table 5.1. Similarly, the evaluation of the ratings of all options on the linguistic scale was divided into five groups, for example, Very_bad [V_B] (0.000, 0.000, 2.500), Bad [B_] (0.000, 2.500, 5.000), Average [A_] (2.500, 5.000, 7.500), Good [G_] (5.000, 7.500, 10.000) and Very_good [V_G] (7.500, 10.000, 10.000). Details of the linguistic scales for evaluating the ratings of the different options are shown in Table 5.2.

The experts DM_r (r = 1......k), used linguistic variables to assess the weightage of the different attributes and the ratings of the various options. Thus, the rth expert provided the $\widetilde{W}_r^j$ weightage to the jth attribute A_j (j = 1......m),. Similarly, r^{th} expert provides the $\widetilde{X}_{ij}^r$ rating to the jth options, OPi (i = 1......n), concerning attributes j. The numbers of steps to apply Fuzzy TOPSIS are described below:

1. The attribute weights and the ratings of options are aggregated here. The mathematical expression for this step may be represented as.

$$\widetilde{wt}j = \frac{1}{k}\left[\widetilde{wt}_j^1 + \widetilde{wt}_j^2 + \ldots + \widetilde{wt}_j^m\right]$$

$$\tilde{x}ij = \frac{1}{k}\left[\tilde{x}_{ij}^1 + \tilde{x}_{ij}^2 + \ldots + \widetilde{x}_{ij}^k\right]$$

Table 5.1: Evaluation of the weights. of the attribute (using a linguistic scale).

Linguistic term	Fuzzy triangular number
"Very_low (V_L)"	"(0.000, 0.000, 0.250)"
"Low (L_)"	"(0.000, 0.250, 0.500)"
"Exactly_equal (E_E)"	"(0.250, 0.500, 0.750)"
"High (H_)"	"(0.500, 0.750, 1.000)"
"Very_high (V_H)"	"(0.750, 1.000, 1.000)"

Table 5.2: Evaluating the ratings of the options (using a linguistic scale).

Linguistic term	Fuzzy triangular number
"Very_Bad (V_B)"	"(0.000, 0.000, 2.500)"
"Bad (B_)"	"(0.000, 2.500, 5.000)"
"Average (A_)"	"(2.500, 5.000, 7.500)"
"Good (G_)"	"(5.000, 7.500, 10.000)"
"Very_good (V_G)"	"(7.500, 10.000, 10.000)"

Source: Junior et. al. (2014)

2. Assemble the attribute matrix, fuzzy decision matrix of options (D˜), and the attributes (()˜). The representation of. the fuzzy decision matrix. is as follows.

$$
\begin{array}{cccccc}
& A1 & A2 & & Aj & Am \\
OP1 & x11 & x12 & x1j & .... & x1m \\
OP2 & x21 & x22 & x2j & .... & x2m \\
D= & . & .... & & & \\
OPn & xn1 & xn2 & xnj & .... & xnm
\end{array}
$$

$$\tilde{W} = [\tilde{w}_1 + \tilde{w}_2 + ... + \tilde{w_m}]$$

3. The fuzzy decision. matrix is normalized by using linear scale transformation for the options ((D)˜). The mathematical expression of normalized fuzzy. decision matrix R˜ is shown as:

$$\tilde{R} = [\tilde{r}ij]mxn$$

$$\tilde{r}ij = (\frac{l_{ij}}{u_j^+}, \frac{m_{ij}}{u_j^+}, \frac{u_{ij}}{u_j^+}) \text{ and } u_j^+ = \max_i u_{ij} \text{(benefit criteria)}$$

$$\tilde{r}ij = \left(\frac{l_j^-}{u_{ij}}, \frac{l_j^-}{m_{ij}}, \frac{u_{ij}}{l_{ij}}\right) \text{ and } l_j^- = \max_i l_{ij} \text{(cost criteria)}$$

4. The weights of the evaluation attribute $\tilde{W}_j$ are multiplied by the elements $\tilde{r}ij$ to obtain a weighted normalized. decision matrix, $\tilde{V}$. The representation of the weighted normalized decision matrix as shown below:

$$\tilde{V} = [\tilde{v}_{ij}]_{mxn} \text{where } \tilde{v}_{ij} \text{is given by the equation}$$
$$\tilde{V}_{ij} = \tilde{X}_{ij} \times \tilde{W}_j$$

5. According to the following equations, the fuzzy positive. ideal solution (FPIS, A+) and the fuzzy negative ideal solution (FNIS, A-), are computed here:

$$A^+ = \{v_1^+, ...v_j^+, ..., v_m^+\}$$

$$A^+ = \{v_1^+, ...v_j^+, ..., v_m^+\}$$

Where $v_j^+ = (1, 1, 1)$ and
$$v_j^- = (0, 0, 0)$$

6. The following equations represent the distances d_j^+ and d_j^- of each alternative from respectively v_j^+ and v_j^- according to

$$d_i^+ = \sum_{j=1}^{n} d_v(v_{ij}, v^+_j)$$

$$d_i^- = \sum_{j=1}^{n} d_v(v_{ij}, v^-_j)$$

Where d_i^+ and d_i^- represents the distance between two fuzzy numbers according to the vertex method. This is expressed as

$$d(\overline{x}, \overline{z}) = \sqrt{\frac{1}{3}[(L_x\text{-}L_z)^2 + (M_x\text{-}M_z)^2 + (U_x\text{-}U_z)^2]}$$

7. In this step, the closeness coefficient, CC_i is calculated according to the given equations:

$$CC_i = \frac{d^-}{d^- + d^+}$$

8. Finally, the ranking is provided to the closeness. coefficient in the decreasing. order. The alternative which has the highest rank is assumed to be best because it is the farthest distance. from the fuzzy negative ideal solution (FNIS).

The flow diagram of methodology has been being given in Figure 5.1.

Application

The target of this study is to select the best selection of suitable strategies in a fuzzy environment. For this, we have identified 11 attributes and 14 options. These 11 attribute are A1: Administration, A2: Consumer, A3: Commodity, A4: Movement, A5: Custom, A6: Workforce, A7: Sovereignty, A8: Technology, A9: Standard of quality, A10: Firm, and A11: Supplementary and the 14 options are as; OP1: Normalization, OP2: Support the learning process and increase motivation, OP3: Segregated decisions, OP4: Reduction in complexity, OP5: Design and organizational structure of work, OP6: Organization of process, OP7: Protection and security issues, OP8: Efficiency of the resources, OP9: An overall infrastructure of broadband for company, OP10: Skilled manpower availability, OP11: Transparency in the information, OP12: Human resource management, OP13: Continuous training and development in the professional skills and OP14: Models for new business. Three experts have been selected from academia and industry to evaluate the weights of attributes and the ratings of the options in the relationship matrix of attributes and options. To execute this study the standard procedure of Fuzzy TOPSIS has been adopted from Junior et. al. (2014). According to Junior et. al. (2014), the weights of attribute in linguistic scale have been classified in to five categories such as, Very_low [V_L] (0.000, 0.000, 0.250), Low [L_] (0.000, 0.250, 0.500), Exactly_equal [E_E]

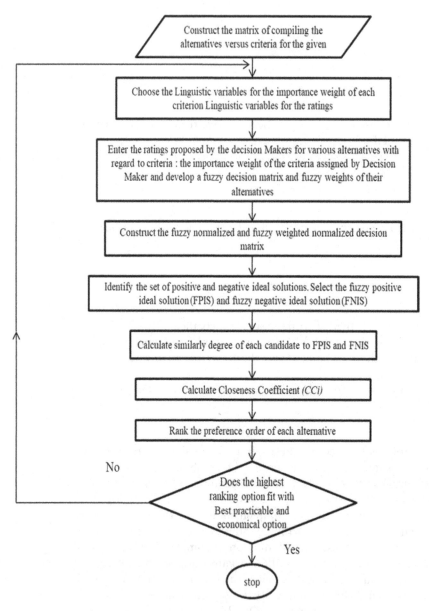

Figure 5.1 Fuzzy TOPSIS method steps and flow

(0.250. 0.500, 0.750), High [H_] (0.500, 0.750, 1.000) and Very_high [V_H] (0.750, 1.000, 1.000). Similarly, the evaluation of the ratings of all options on the linguistic scale was divided into five groups, for example, Very_bad [V_B] (0.000, 0.000, 2.500), Bad [B_] (0.000, 2.500, 5.000), Average [A_] (2.500, 5.000, 7.500), Good [G_] (5.000, 7.500, 10.000) and Very_good [V_G] (7.500, 10.000, 10.000). The responses of weights of attributes are recorded by the three experts are as shown in Table 5.3:

Table 5.4 shows the aggregate weight of all attributes which is actually the average of all weights provided by all 3 experts to the attributes.

Table 5.5 represent the ratings of all options corresponding to all attributes given by the first expert. In the same way, Table 5.6 and 5.7 provides the rating of all options corresponding to all attributes given by the second and third expert.

After finding the aggregate weight and aggregate rating, normalized decision. matrix, weighted. normalized decision matrix, fuzzy positive. ideal solution (FPIS, A$^+$), fuzzy negative ideal. solution (FNIS, A$^-$), the distances d_j^+ and d_j^- of each alternative. from FPIS. and FNIS. respectively, and the closeness coefficient is calculated. The closeness. coefficients data is shown in Table 5.9.

From. Table 5.9, It is clear. that the most selection of suitable strategy strategies for creating the roadmap for Industry-4.) are OP10: Skilled manpower availability, OP9: An overall infrastructure of broadband for company, OP14: Models for new business, OP8: Efficiency of the resources and OP7: Protection and security issues.

Conclusion

This study shows that the fuzzy TOPSIS approach is widely accepted, choosing the most. appropriate strategy. for Industry. 4.0 (I4.0). Relevant attributes and strategies are determined by expert opinion and literature reviews. In the study, three experts were selected to rate 14 options of the 11 attributes. The subjective judgment of the expert evaluates the relative. importance of the various attributes and the selection scores of those attributes are a limitation of the study. According to experts, 'A8: Technology' is one of the most important attributes, while choosing the right strategy for strategic I4.0. Based on the results obtained, "Skilled manpower availability" can be treated as the premier option as the selection. of suitable strategy for I4.0. Also, "an overall infrastructure of broadband for the company", "models for new business", "efficiency of the resources" and "protection and security issues" are determined as the second, third, fourth and the fifth important strategy to follow. It is believed that this. study may be a road. map for selecting the appropriate strategy for I4.0. Different versions of the. fuzzy set can be used for a given suggested method, or for implementing different types of MADM methods for the property evaluation and rating of options, rather than providing fuzzy TOPSIS for future research.

References

Chen, C. T. (2000). Extensions of the TOPSIS for group decision-making under fuzzy environment. *Fuzzy sets and systems*, 114(1), 1-9.

Chen, S.-M. and Huang, Z.-C. (2017). Multi-attribute decision making based on interval-valued intuitionistic fuzzy values and particle swarm optimization techniques. *Information Sciences*, 397, 206–218.

Condry, M. W. and Nelson, C. B. (2016). Using smart edge IoT devices for safer, rapid response with industry IoT control operations. *Proceedings of the IEEE*, 104(5), 938–946.

Erdogan, M., Ozkan, B., Karasan, A., and Kaya, I. (2018). Selecting the best strategy for Industry-4.o applications with a case study. *Industrial Engineering in the Industry 4.0 Era*, 109–119.

Hermann, M., Pentek, T., and Otto, B. (2016). Design principles for Industrie 4.0 scenarios. *In Proceedings of the Annual Hawaii International Conference on System Sciences*, (pp. 3928–3937).

Junior, F. R. L., Osiro, L., and Carpinetti, L. C. R. (2014). A comparison between Fuzzy AHP and Fuzzy TOPSIS methods to supplier selection. *Applied Soft Computing*, 21, 194–209.

Kagermann, H., Wahlster, W., and Helbig, J. (2013). Recommendations for implementing the strategic initiative industrie 4.0 April 2013 securing the future of German manufacturing industry. Final report of the Industrie 4.0 Working Group Accessed at https://www.din.de/blob/76902/e8cac883f42bf28536e7e8165993f1fd/recommendations-for-implementing-industry-4-0-data.pdf.

Kaya, İ, Erdoğan, M., Karaşan, A., and Özkan, B. (2020). Creating a road map for Industry-4.o by using an integrated fuzzy multi-attribute decision-making methodology. *Soft Computing*, 24, 17931–17956.

Lasi, H., Fettke, P., Kemper, H.-G., Feld, T., and Hoffmann, M. (2014). Industry-4.o. *Business and Information Systems Engineering*, 6(4), 239–242. https://doi.org/10.1007/s12599-014-0334-4.

Luthra, S. and Mangla, S. K. (2018). Evaluating challenges to industry-4.o initiatives for supply chain sustainability in emerging economies. *Process Safety and Environmental Protection*, 117, 168–179.

Medic, N., Anisˇicˊ, Z., Lalicˊ, B., Marjanovicˊ, U., and Brezocnik, M. (2019). Hybrid fuzzy multi-attribute decision making model for evaluation of advanced digital technologies in manufacturing: Industry-4.o perspective. *Advances in Production Engineering & Management*, 14(4), 483–493. https://doi.org/10.14743/apem2019.4.343.

Moktadir, M. A., Dwivedi, A., Khan, N. S., Paul, S. K., Khan, S. A., Ahmed, S., and Sultana, R. (2021). Analysis of risk factors in sustainable supply chain management in an emerging economy of leather industry. *Journal of Cleaner Commodityion*, 283, 124641.

Narula, S., Kumar, A., Prakash, S., Dwivedi, M., Puppala, H., and Talwar, V. (2021). Modelling and analysis of challenges for industry-4.o implementation in medical device industry to post Covid-19 scenario. *International Journal of Supply and Operations Management*, 10(2), 117–135.

Prakash, S., Jasti, N. V. K., Chan, F. T. S., Nilaish, Sharma, V. P., and Sharma, L. K. (2021). Decision modelling of critical success factors for cold chains using the DEMATEL approach: a case study. *Measuring Business Excellence*, 26(3), 263–287. https://doi.org/10.1108/MBE-07-2020-0104.

Priya, B. and Malhotra, J. (2019). 5GAuNetS: an autonomous 5G network selection framework for industry-4.o. *Soft Computing*, 24(13), 9507–9523. https://doi.org/10.1007/s00500-019-04460-y.

Qin, J., Liu, Y., and Grosvenor, R. (2016). A categorical framework of manufacturing for industry-4.o and beyond. *Procedia CIRP*, 52, 173–178.Ramos, L., Loures, E., Deschamps, F., and Venaˆncio, A. (2020). Systems evaluation methodology to attend the digital projects requirements for industry-4.o. *International Journal of Computer Integrated Manufacturing*, 33(4), 398–410.

Schla¨pfer, R. C., Koch, M., and Marketer, P. (2015). Industry-4.o Challenges and Solutions for the Digital Transformation and use of Exponential Technologies. Deloitte, Zurique.

Schumacher, A., Erol, S., and Sihn, W. (2016). A maturity model for assessing Industry-4.o readiness and maturity of manufacturing enterprises. *Procedia CIRP*, 52(1), 161–166.

Segovia, V. R. and Theorin, A. (2012). History of Control History of PLC and DCS. University of Lund.

Simon, J., Trojanova, M., Zbihlej, J., and Sarosi, J. (2018). Mass customization model in food industry using Industry-4.o standard with fuzzy based multi-attribute decision making methodology. *Advances in Mechanical Engineering*, 10(3), 1–10. 1687814018766776.

Sung, T. K. (2017). Industry-4.o: A Korea perspective. *Technological Forecasting and Social Change*, 132, 1–6.

Tzeng, G.-H. and Huang, J.-J. (2011). Multiple Attribute Decision Making: Methods and Applications. CRC Press, Cambridge.

Womack, J. P., Jones, D. T., and Roos, D. (1990). The Machine that Changed the World. Free Press, New York.

Yang, H., Chen, F., and Aliyu, S. (2017). Modern software cybernetics: new trends. *Journal of Systems and Software*, 124, 169–186.

Appendix

Table 5.3: Response of the three experts about the weight of attribute

DMs	A1	A2	A3	A4	A5	A6	A7	A8	A9	A10	A11
DM1	V_H	H_	H_	H_	E_E	V_H	L_	V_H	V_H	V_H	E_E
DM2	H_	V_H	V_H	H_	E_E	L_	H_	V_H	V_H	E_E	E_E
DM3	H_	H_	H_	H_	V_H	H_	V_H	V_H	H_	H_	V_L

The aggregation of all attributes in linguistic scale can be found out by calculating the means of all three decision. The aggregated value of weights of attribute are as follow in Table 5.4:

Table 5.4: Aggregated weight of attribute.

A1	A2	A3	A4	A5	A6	A7	A8	A9	A10	A11
(0.583, 0.833, 1.0)	(0.583, 0.833, 1.0)	(0.583, 0.833, 1.0)	(0.50, 0.75, 1.0)	(0.417, 0.833, 0.833)	(0.417, 0.667, 0.833)	(0.417, 0.583, 0.833)	(0.75, 1.0, 1.0)	(0.667, 0.917, 1.0)	(0.5, 0.75, 0.917)	(0.167, 0.333, 0.583)

The responses of ratings of options recorded by the three decisions are shown in the following Tables 5.5-5.7:

Table 5.5: Ratings of alternative by first expert.

	A1	A2	A3	A4	A5	A6	A7	A8	A9	A10	A11
OP1	A_	G_	V_G	G_	A_	B_	G_	V_G	G_	G_	A_
OP2	V_G	G_	G_	V_G	A_	G_	A_	V_G	G_	V_G	G_
OP3	V_G	A_	G_	V_G	B_	G_	A_	V_G	A_	V_G	A_
OP4	G_	A_	G_	G_	B_	A_	A_	V_G	G_	B_	G_
OP5	A_	A_	G_	G_	A_	V_G	G_	G_	A_	V_G	G_
OP6	G_	G_	A_	G_	G_	G_	V_G	G_	G_	G_	G_
OP7	G_	A_	A_	G_	G_	G_	G_	G_	V_G	G_	A_
OP8	G_	A_	V_G	G_	G_	G_	G_	G_	V_G	G_	A_
OP9	G_	A_	V_G	G_	A_	G_	G_	V_G	G_	G_	G_
OP10	V_G	A_	G_	G_	A_	V_G	A_	G_	G_	G_	G_
OP11	G_	V_G	G_	A_	V_G	V_G	G_	G_	G_	A_	G_
OP12	A_	G_	G_	A_	V_G	V_G	G_	G_	A_	A_	G_
OP13	G_	G_	V_G	A_	V_G	G_	A_	V_G	G_	G_	V_G
OP14	G_	A_	G_	G_	A_	G_	A_	V_G	G_	G_	G_

Table 5.6: Ratings of alternative by second expert.

	A1	A2	A3	A4	A5	A6	A7	A8	A9	A10	A11
OP1	A_	A_	G_	A_	A_	A_	A_	V_G	A_	A_	A_
OP2	A_	V_G	G_	A_	A_	A_	A_	A_	A_	V_G	A_
OP3	G_	A_	A_	G_	A_	A_	G_	V_G	G_	G_	A_
OP4	G_	A_	A_	B_	G_	A_	A_	V_G	G_	G_	G_
OP5	G_	B_	V_G	A_	A_	A_	G_	V_G	G_	G_	A_
OP6	A_	B_	V_G	A_	B_	G_	A_	V_G	V_G	G_	A_
OP7	V_G	V_G	V_G	G_	G_	A_	G_	V_G	G_	G_	V_G
OP8	B_	A_	V_G	V_G	A_	G_	V_G	V_G	G_	G_	V_G
OP9	V_G	V_G	V_G	A_	G_	A_	G_	V_G	V_G	G_	V_G
OP10	A_	V_	G_	V_G	G_	A_	A_	V_G	G_	V_G	A_
OP11	V_G	V_G	B_	A_	A_	G_	G_	A_	V_G	G_	A_
OP12	A_	A_	V_	G_	G_	A_	A_	V_G	G_	G_	A_
OP13	V_G	V_G	V_G	G_	A_	G_	G_	V_G	G_	G_	V_G
OP14	V_G	A_	G_	A_	A_	G_	G_	V_G	G_	G_	V_G

Table 5.7: Ratings of alternative by third expert.

	A1	A2	A3	A4	A5	A6	A7	A8	A9	A10	A11
OP1	G_	B_	A_	A_	G_	G_	B_	A_	A_	G_	A_
OP2	A_	G_	G_	G_	A_	A_	A_	B_	G_	A_	B_
OP3	A_	G_	A_	A_	A_	G_	G_	G_	A_	G_	A_
OP4	G_	V_G	G_	V_G	G_	A_	A_	A_	G_	A_	G_
OP5	A_	G_	V_G	A_	A_	A_	G_	V_G	A_	G_	G_
OP6	G_	A_	V_G	G_	V_G	G_	A_	V_G	G_	A_	A_
OP7	A_	G_	G_	G_	G_	G_	G_	G_	A_	A_	V_G
OP8	A_	A_	G_	G_	G_	V_G	V_G	G_	A_	V_G	G_
OP9	G_	A_	A_	A_	A_	G_	A_	A_	G_	G_	G_
OP10	V_G	G_	A_	A_	V_G	V_G	G_	G_	V_G	G_	V_G
OP11	V_G	V_G	G_	V_G	G_	G_	B_	B_	G_	G_	A_
OP12	A_	A_	G_	G_	V_G	A_	V_G	G_	G_	A_	G_
OP13	A_	V_G	A_	G_	G_	B_	A_	B_	V_G	B_	A_
OP14	V_G	G_	G_	V_G	A_	G_	G_	A_	G_	G_	G_

The aggregated ratings of the options concerning the attribute are provided in Table 5.8.

Table 5.8: Aggregated ratings of options.

	A1	A2	A3	A4	A5	A6	A7	A8	A9	A10	A11
OP1	(3.33, 5.83, 8.33)	(2.5, 5, 7.5)	(7.5, 7.5, 9.16)	(7.5, 7.5, 9.16)	(2.5, 5, 7.5)	(2.5, 5.0, 7.5)	(2.5, 5, 7.5)	(3.33, 5, 6.67)	(3.33, 5.83, 8.33)	(4.16, 6.67, 9.16)	(2.5, 5.0, 7.5)
OP2	(4.17, 6.67, 8.33)	(5.83, 8.33, 10)	(5, 7.5, 10)	(5, 7.5, 9.16)	(2.5, 5, 7.5)	(3.33, 5.83, 8.33)	(2.5, 5.0, 7.5)	(3.33, 5.83, 7.5)	(4.16, 6.67, 9.16)	(5.83, 8.33, 9.16)	(2.5, 5, 7.5)
OP3	(5.83, 7.5, 9.16)	(3.33, 5.83, 8.33)	(3.33, 5.83, 8.33)	(4.16, 6.66, 8.33)	(1.67, 4.16, 5.0)	(4.16, 6.66, 9.16)	(4.16, 6.66, 9.16)	(6.67, 9.16, 10)	(3.33, 5.83, 8.33)	(5.83, 8.33, 10.0)	(2.5, 5.0, 7.5)
OP4	(5, 7.5, 10)	(4.16, 5.83, 8.33)	(4.16, 6.66, 9.16)	(5.83, 8.33, 10)	(3.33, 5.83, 8.33)	(2.5, 5.0, 7.5)	(2.5, 5.0, 7.5)	(5.83, 8.33, 9.16)	(5.0, 7.5, 10.0)	(2.5, 5, 7.5)	(5.0, 7.5, 10.0)
OP5	(3.33, 5.83, 8.33)	(2.5, 5, 7.5)	(6.66, 9.16, 10)	(2.5, 5.0, 7.5)	(2.5, 5.0, 7.5)	(4.16, 6.66, 8.33)	(5.0, 7.5, 10.0)	(6.67, 9.16, 10.0)	(3.33, 5.83, 8.33)	(5.83, 8.33, 10.0)	(4.16, 6.67, 9.16)
OP6	(4.16, 6.67, 9.16)	(2.5, 5, 7.5)	(5.83, 8.33, 9.16)	(4.16, 6.66, 9.16)	(4.16, 6.67, 8.33)	(5.0, 7.5, 10.0)	(4.16, 6.66, 8.33)	(6.66, 9.16, 10.0)	(5.83, 8.33, 10.0)	(4.16, 6.67, 9.16)	(3.33, 5.83, 8.33)
OP7	(5, 7.5, 9.16)	(5, 7.5, 9.16)	(5, 7.5, 9.16)	(4.16, 6.66, 9.16)	(5.0, 7.5, 10.0)	(4.16, 6.66, 9.16)	(5.0, 7.5, 10.0)	(5.83, 8.33, 10.0)	(5.0, 7.5, 9.16)	(4.16, 6.67, 9.16)	(5.83, 8.33, 9.16)
OP8	(2.5, 5, 7.5)	(2.5, 5, 7.5)	(6.66, 9.16, 10)	(5, 7.5, 10)	(4.16, 6.66, 9.16)	(5.83, 8.33, 10.0)	(6.66, 9.16, 10.0)	(5.83, 8.33, 10.0)	(5.0, 7.5, 9.16)	(5.83, 8.33, 10.0)	(5.0, 8.33, 9.16)
OP9	(5.83, 8.33, 10.0)	(4.16, 6.66, 8.33)	(5.83, 8.33, 9.16)	(5, 7.5, 9.16)	(3.33, 5.83, 8.33)	(4.16, 6.66, 9.16)	(4.16, 6.67, 9.16)	(5.83, 8.33, 9.16)	(5.83, 8.33, 10.0)	(5.0, 7.5, 10.0)	(5.83, 8.33, 10.0)
OP10	(5.83, 8.33, 9.16)	(5, 7.5, 9.16)	(4.16, 6.66, 9.16)	(3.33, 5.83, 8.33)	(5.0, 7.5, 9.16)	(5.83, 8.33, 9.16)	(3.33, 5.83, 8.33)	(5.83, 8.33, 10.0)	(5.83, 8.33, 10.0)	(5.83, 8.33, 10.0)	(5.0, 7.5, 9.16)
OP11	(6.67, 9.16, 10)	(7.5, 10, 10)	(3.33, 5.83, 8.33)	(5.83, 8.33, 9.16)	(5, 7.5, 9.16)	(6.66, 10.0, 10.0)	(3.33, 5.83, 8.33)	(2.5, 5, 7.5)	(5.83, 8.33, 10.0)	(4.16, 6.67, 9.16)	(3.33, 5.83, 8.33)
OP12	(2.5, 2.5, 7.5)	(3.33, 5.83, 8.33)	(5.83, 9.16, 10)	(3.33, 5.83, 8.33)	(6.66, 9.16, 10)	(4.16, 6.66, 8.33)	(5, 7.5, 9.16)	(5.83, 9.16, 10)	(4.16, 6.67, 9.16)	(3.33, 5.83, 8.33)	(4.16, 6.67, 9.16)
OP13	(5, 7.5, 9.16)	(6.66, 9.16, 10)	(5.83, 8.33, 9.16)	(4.16, 6.66, 9.16)	(5, 7.5, 9.16)	(3.33, 5.83, 8.33)	(3.33, 5.83, 8.33)	(5.0, 7.5, 8.33)	(5.83, 8.33, 10.0)	(3.33, 5.83, 8.33)	(5.83, 8.33, 9.16)
OP14	(6.67, 9.16, 10)	(3.33, 5.83, 8.33)	(5, 7.5, 10)	(5.83, 8.33, 10)	(2.5, 5.0, 7.5)	(5.0, 7.5, 10.0)	(4.16, 6.66, 9.16)	(5.83, 8.33, 9.16)	(5.0, 7.5, 10.0)	(5.0, 7.5, 10.0)	(5.83, 9.16, 10.0)

Table 5.9: Closeness coefficient of each alternative.

	Options	CC	%CC	Ranking
OP1	Normalization	0.47	46.51	
OP2	Support the learning process and increase motivation	0.51	50.97	
OP3	Segregated decisions	0.51	51.10	
OP4	Reduction in complexity	0.51	50.70	
OP5	Design and organizational structure of work	0.52	51.74	
OP6	Organization of process	0.53	53.45	
OP7	Protection and security issues	0.55	54.78	5
OP8	The efficiency of the resources	0.55	55.06	4
OP9	An overall infrastructure of broadband for the company	0.55	55.20	2
OP10	Skilled manpower availability	0.55	55.30	1
OP11	Transparency in the information	0.55	54.65	
OP12	Human resource management	0.52	52.13	
OP13	Continuous training and development in the professional skills	0.54	53.70	
OP14	Models for new business	0.55	55.17	3

6 A novel approach to population growth model via Laplace decomposition method

Deepika Jain and Alok Bhargava[a]

Department of Mathematics & Statistics, Manipal University Jaipur, India

Abstract

Population growth and its consequences is one of the major problems of our world. Based on mathematical population growth model, Malthus has given the theory of population in arithmetic manner which was correspond to exponential growth and food supply growth model. Later, more realistic model was developed by Belgian mathematician P. F. Verhulst also known as logistic growth model. This model involves non-linear ordinary differential equation (ODE). Due to the significance of nonlinear ODEs, several authors have given the solution of the model by different techniques. In our work, we employ the 'Laplace decomposition method' (LDM) for finding the result of the Verhulst's model. The graphical interpretation of the behavior of result is also mentioned.

Keywords: Adomian decomposition, Laplace decomposition method, Laplace transform, non-linear ordinary differential equation, population growth model

Introduction

From the last few decades, many researchers are attracted to receiving analytical and numerical solutions to non-linear differential equations due to their wide-ranging applications in many fields of applied and pure mathematics, engineering, computational sciences, physical and biological sciences, etc. So, it is very significant to be aware of all the mathematical techniques that will provide the solution of the problem in a rapid and efficient way. In this episode, we exhibit the approach of Laplace decomposition method (LDM) to a nonlinear model of population growth given by Verhulst (1838), which was the extension of the model given by Malthus. The Malthusian theory of population is related to exponential population growth and food supply growth in an arithmetic manner T. R. Malthus proposed this theory in 1798, which was based on the population principle (Ramirez-Cando et. al., 2018).

The Malthusian growth model was described by the differential equation:

$$\frac{dN(t)}{dt} = P'^{*}N(t), \qquad (1.1)$$

(t): Population size, P'^{*}: Malthusian parameter

[a]alok.bhargava@jaipur.manipal.edu

DOI: 10.1201/9781003450917-6

This model was having some limitations due to environmental restrictions. Later, by the relevant studies it was found that the more realistic model is logistic model (Ausloos, 2006; Ramirez-Cando et. al., 2018; Alkahtani et. al., 2017), which was developed by Belgian mathematician (Verhulst, 1838). In this model he suggested that the rate of population increase may be limited and further he describes the model by the non-linear differential equation

$$\frac{dN(t)}{dt} = P''N(t)\left[1 - \frac{N(t)}{K'}\right] \qquad (1.2)$$

(t): Population size, P'^*: Rate of maximum population growth, and K': Carrying capacity.

For a single species, such as the human population, the population of rabbits, the population of an endangered species, the population of microbes, etc., this continuous population model is used. When it comes to understanding the dynamic process at play and creating useful predictions, the Verhulst model is particularly helpful. This model is also having applications in statistics and machine learning, medicine, Fermi-Dirac distribution, reaction models, material science, Linguistic, agriculture, economics, and sociology (Homan, 2010; Pomeranz, 2000; Rocha et. al., 2017; Yin and Zelenay, 2018) etc. Looking such applications of the model in diverse fields, the solution of the model for different fields was given by some authors with different approaches. In this episode, obtaining the Verhulst population growth model's solution is the primary goal of the study we have provided here by a new and powerful technique LDM (Khuri, 2001). Yusufolu (2006) applied for the Duffing equation, while Khuri (2001) used for the approximate solution of a class of nonlinear ordinary differential equations, Elgazery (2008) uses for Falkner-Skan equation, Hosseinzadeh et. al. (2010) uses Klein-Gordon equation, Khan and Faraz (2011) uses for boundary layer equations, Rani and Mishra (2019) uses for Volterra integral and integro-differential equations, etc. Also, for more recent work one can refer (Khan, et. al., 2020; Ogunsola et. al., 2022; Shah et. al., 2019).

Laplace decomposition method

The LDM, which combines the Adomian decomposition method and the Laplace transform, illustrates how the Laplace transform can be used to approximation the solutions of the nonlinear differential equations.

The LDM is a mixture of Laplace transform and Adomian decomposition method and exhibit how the Laplace transform may be used to approximate the solutions of the nonlinear differential equations by combining with Adomian's decomposition technique (Adomian, 1990).

The popular Laplace transform (Schiff, 1999) is described as

$$L[f(t)] = \int_0^\infty e^{-st} f(t)dt, \qquad (1.3)$$

The Laplace transform is applied to the differential equation being considered, assuming that the solution of the differential equation is of the form $= \sum_{n=0}^{\infty} u_n$. Additionally, the nonlinear term should be broken down using Adomian polynomials (Adomian, 1990) and then by constructing an iterative sequence, determine all the u_n; $n = 0,1,\ldots,\infty$. The method's significance stems from how quickly the solution converges to the exact solution. LDM is devoid of round-off mistakes and does not involve any minor or significant parameters. It has therefore been given more weight than other approximation techniques. like other analytical methods, there is no need for linearization and discretization in LDM thus, results obtained by LDM show its robustness.

To illustrate the LDM (Ayata and Ozkan, 2020; Khuri, 2001) let us take the following non-linear differential equation into consideration:

$$DN(t) + P'N(t) + QN(t) = f(t), \tag{1.4}$$

$$N(t) = N(0) \ at \ t{=}0 \tag{1.5}$$

where $QN(t)$ is the nonlinear term, D is the highest order derivative, P' is a linear differential operator of the order smaller than D. The standard interpretation of N(t) applies.

Using equation (1.4) and the Laplace transform, we have

$$L[DN(t) + P' N(t) + QN(t)] = L[f(t)]$$
$$L[N(t)] = \frac{1}{s}N(0) - \frac{1}{s}L[P' N(t) + QN(t)] + L[f(t)] \tag{1.6}$$

taking inverse Laplace transform we get

$$N(t) = L^{-1}\left\{\frac{1}{s}N(0) - \frac{1}{s}L[P' N(t) + QN(t)] + L[f(t)]\right\} \tag{1.7}$$

Now as per the next step of LDM, we represent the solution of (1.7) in an infinite series as

$$N(t) = \sum_{i=0}^{\infty} N_i(t) = N_0(t) + N_1(t) + N_2(t) + \cdots + N_i(t) + \cdots \tag{1.8}$$

The nonlinear operator is decomposed as

$$QN(t) = \sum_{i=0}^{\infty} A_i \tag{1.9}$$

A_i are Adomian polynomials, which are described as

$$A_i = \frac{1}{i!}\left[\frac{d^i}{d\varepsilon^i}\{Q \sum_{i=0}^{\infty}(\varepsilon^i N_i)\}\right]_{\varepsilon=0}, \quad i = 0,1,2,\ldots, \tag{1.10}$$

putting (1.8) and (1.9) in (1.7), we find

$$N(t) = N(0) - L^{-1}\left\{\frac{1}{s}L[P' N(t) + QN(t)]\right\} \tag{1.11}$$

$$N_0(t) = N(0),$$

$$N_1(t) = -L^{-1}\left\{\frac{1}{s}L[P' N_0(t) + A_0]\right\}, \tag{1.12}$$

$$N_{i+1}(t) = -L^{-1}\left\{\frac{1}{s}L[P' N_i(t) + A_i]\right\}, \ i \geq 1 \ . \tag{1.13}$$

Further by using (1.9) and (1.10), the solution of (1.4) can be obtained easily.

Verhulst population growth model and LDM

Here, in this section we will find the solution of VPGM by LDM.
  The VPGM described by the following mathematical equation:

$$\frac{dN(t)}{dt} = P'N(t)\left[1 - \frac{N(t)}{K'}\right] \tag{2.1}$$

and its solution is

$$N(t) = \frac{K'}{1 + \left(\frac{K'}{N(0)} - 1\right)e^{-P't}} \tag{2.2}$$

where $N(t)$, P' and K' has the same meaning (as mentioned with (1.2)) and $N(0)$ is the size of the starting population.
  Solution: According to the methodology of LDM, we first apply the Laplace transform (Schiff, 1999) on (2.1), then, it gives

$$L\{N(t)\} = \frac{N(0)}{s-P'} - \frac{P'}{K'(s-P')}L\{N^2(t)\} \tag{2.3}$$

By taking the inverse Laplace Transform of (2.3)

$$N(t) = N(0)e^{P't} - \frac{P'}{K'}L^{-1}\left[\frac{1}{(s-P')}L\{N^2(t)\}\right] \tag{2.4}$$

Now as per the next step of LDM, we represent the solution of (2.1) in an infinite series as

$$N(t) = \sum_{i=0}^{\infty} N_i(t) = N_0(t) + N_1(t) + N_2(t) + \cdots + N_i(t) + \cdots \tag{2.5}$$

And for obtaining the Adomian decomposition, let us set

$$L\{N^2(t)\} = L\left\{\sum_{i=0}^{\infty} A_i(t)\right\} \tag{2.6}$$

where $A_i(t)$ are Adomian Polynomials which depends upon $N_0(t), N_1(t), N_2(t),\ldots$ where

$$A_0(t) = N_0^2(t) \tag{2.7}$$

$$A_1(t) = N_1(t)\frac{d}{dt}\left(N_0^2(t)\right) = 2N_0(t)N_1(t) \tag{2.8}$$

$$A_2(t) = N_2(t)\frac{d}{dt}\left(N_0^2(t)\right) + \frac{1}{2}N_1^2(t)\frac{d^2}{dt^2}\left(N_0^2(t)\right) = 2N_0(t)N_2(t) + N_1^2(t) \tag{2.9}$$

Similarly, other Adomian polynomials $A_3(t), A_4(t), \ldots$ can be obtained easily. Now using (2.6) in (2.4), we have

$$N(t) = N(0)e^{P't} - \frac{P'}{K'}L^{-1}\left[\frac{1}{(s-P')}L\{\textstyle\sum_{i=0}^{\infty} A_i(t)\}\right] = N(0)e^{P't} - \frac{P'}{K'}L^{-1}\left[\frac{1}{(s-P')}L\{A_0(t) + A_1(t) + A_2(t) + \cdots\}\right] \tag{2.10}$$

Comparing (2.4) and (2.9), we get the values of $N_0(t), N_1(t), N_2(t),\ldots$ as

$$N_0(t) = N(0)e^{P't} \tag{2.11}$$

$$N_1(t) = -\frac{P'}{K'}L^{-1}\left[\frac{1}{(s-P')}L\{A_0(t)\}\right] = -\frac{P'}{K'}L^{-1}\left[\frac{1}{(s-P')}L\{\{N(0)\}^2 e^{2P't}\}\right]$$
$$= -\frac{P'\{N(0)\}^2}{K'}L^{-1}\left[\frac{1}{(s-P')(s-2P')}\right] = \frac{\{N(0)\}^2}{K'}e^{P't}\left(1 - e^{P't}\right) \tag{2.12}$$

$$N_2(t) = -\frac{P'}{K'}L^{-1}\left[\frac{1}{(s-P')}L\{A_1(t)\}\right] = -\frac{P'}{K'}L^{-1}\left[\frac{1}{(s-P')}L\{2N_0(t)N_1(t)\}\right]$$
$$= -\frac{2P'}{K'}L^{-1}\left[\frac{1}{(s-P')}L\left\{\frac{\{N(0)\}^3}{K'}e^{2P't}\left(1 - e^{P't}\right)\right\}\right] = -\frac{\{N(0)\}^3}{K'^2}L^{-1}\left[\frac{1}{(s-P')}\left\{\frac{1}{(s-2P')} - \frac{1}{(s-3P')}\right\}\right]$$
$$\Rightarrow N_2(t) = \frac{\{N(0)\}^3}{K'^2}e^{P't}\left(1 - e^{P't}\right)^2 \tag{2.13}$$

$$N_3(t) = -\frac{P'}{K'}L^{-1}\left[\frac{1}{(s-P')}L\{A_2(t)\}\right] = -\frac{P'}{K'}L^{-1}\left[\frac{1}{(s-P')}L\{2N_0(t)N_2(t) + N_1^2(t)\}\right] =$$
$$-\frac{P'}{K'}L^{-1}\left[\frac{1}{(s-P')}L\left\{\frac{2\{N(0)\}^4}{K'^2}e^{2P't}\left(1 - e^{P't}\right)^2 + \frac{\{N(0)\}^4}{K'^2}e^{2P't}\left(1 - e^{P't}\right)^2\right\}\right]$$
$$-\frac{3P'\{N(0)\}^4}{K'^3}L^{-1}\left[\frac{1}{(s-P')}L\left\{e^{2P't}\left(1 - e^{P't}\right)^2\right\}\right]$$
$$= -\frac{3P'\{N(0)\}^4}{K'^3}L^{-1}\left[\frac{1}{(s-P')}\left\{\frac{1}{(s-2P')} - \frac{2}{(s-3P')} + \frac{1}{(s-4P')}\right\}\right]$$
$$\Rightarrow N_3(t) = \frac{\{N(0)\}^4}{K'^2}e^{P't}\left(1 - e^{P't}\right)^3 \tag{2.14}$$

Similarly

$$N_4(t) = \frac{\{N(0)\}^5}{K'^4} e^{P't} \left(1 - e^{P't}\right)^4$$

$$N_5(t) = \frac{\{N(0)\}^6}{K'^5} e^{P't} \left(1 - e^{P't}\right)^5, \dots \tag{2.15}$$

Now in view of (2.11), (2.12), (2.13), (2.14) and (2.15), from (2.5), we have

$$N(t) = N(0)e^{P't} + \frac{\{N(0)\}^2}{K'} e^{P't} \left(1 - e^{P't}\right) + \frac{\{N(0)\}^3}{K'^2} e^{P't}\left(1 - e^{P't}\right)^2 + \frac{\{N(0)\}^4}{K'^3} e^{P't}\left(1 - e^{P't}\right)^3 + \frac{\{N(0)\}^5}{K'^4} e^{P't}\left(1 - e^{P't}\right)^4 + \frac{\{N(0)\}^6}{K'^5} e^{P't}\left(1 - e^{P't}\right)^5 + \cdots$$

$$= N(0)e^{P't}\left[1 + \frac{N(0)}{K'}\left(1 - e^{P't}\right) + \frac{\{N(0)\}^2}{K'^2}\left(1 - e^{P't}\right)^2 + \frac{\{N(0)\}^3}{K'^3}\left(1 - e^{P't}\right)^3 + \frac{\{N(0)\}^4}{K'^4}\left(1 - e^{P't}\right)^4 + \frac{\{N(0)\}^5}{K'^5}\left(1 - e^{P't}\right)^5 + \cdots\right]$$

$$= N(0)e^{P't}\left[1 - \frac{N(0)}{K'}\left(1 - e^{P't}\right)\right]^{-1} = \frac{N(0)e^{P't}}{\left[1 - \frac{N(0)}{K'}\left(1 - e^{P't}\right)\right]}$$

From which the desired result can be achieved easily.

Numerical illustration

In this part we found a record for numerical explanation of population growth model (presented in Table 6.1 and Table 6.2) for $0 < t < 1$, $P' > 0$ and $P' < 0$. We plot the graphs for (2.2). In each graph, putting different values to the parameters. We take $K' = 1$, $N(0) = 0.5$ and $P' > 0$ in Figure 6.1. Similarly, Figure 6.2 plotted by taking $K' = 1$, $N(0) = 0.5$ and $P' < 0$. It is quite simple to study and observe the characteristics of the solutions for various parameters and time intervals.

Table 6.1: Numerical values of $N(t)$ for variables t and $P' > 0$

t	$P' = 0.10$	$P' = 0.30$	$P' = 0.50$	$P' = 0.70$	$P' = 0.90$
1	0.524979	0.574442	0.622459	0.668187	0.710949
2	0.549833	0.645656	0.731058	0.802183	0.858148
3	0.574442	0.710949	0.817574	0.890903	0.937026
4	0.598687	0.768524	0.880797	0.942675	0.973403
5	0.622459	0.817574	0.924141	0.970687	0.989013
6	0.645656	0.858148	0.952574	0.985225	0.995503
7	0.668187	0.890903	0.970687	0.985225	0.998167
8	0.689974	0.916827	0.982013	0.996315	0.999253
9	0.710949	0.937026	0.989013	0.998167	0.999696
10	0.731058	0.952574	0.993307	0.999088	0.999876

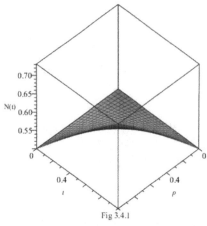

Fig 3.4.1

Figure 6.1 The behavior of $N(t)$ w.r.t t and P'

Table 6.2: Numerical values of $N(t)$ for variables t and $P' > 0$

t	$P' = -0.10$	$P' = -0.30$	$P' = -0.50$	$P' = -0.70$	$P' = -0.90$
1	0.475020	0.425557	0.377540	0.331812	0.289050
2	0.450166	0.354343	0.268941	0.197816	0.141851
3	0.425557	0.289050	0.182425	0.109096	0.062973
4	0.401312	0.231475	0.119202	0.057324	0.026596
5	0.377540	0.182425	0.075858	0.029312	0.010986
6	0.354343	0.141851	0.047425	0.014774	0.004496
7	0.331812	0.109096	0.029312	0.007391	0.001832
8	0.310025	0.083172	0.017986	0.003684	0.000746
9	0.289050	0.062973	0.010986	0.001832	0.000303
10	0.268941	0.047425	0.006692	0.000911	0.000123

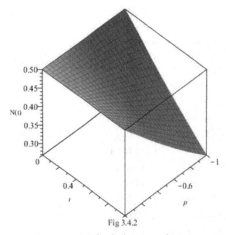

Fig 3.4.2

Figure 6.2 The behavior of $N(t)$ w.r.t t and P'

Conclusion

In the present work, the approach of the Laplace decomposition method (LDM) is considered to find the solution of VPGM. As in the LDM, the discretization or restrictive assumptions are not taken in consideration, because there are no round-off errors in the results, the numerical computations are significantly reduced. Hence due the strength of the technique, the solution obtained strongly validate the results obtained by other methods. The work presented here provides the opportunity to researchers to apply the method and the model in sensitive analysis of results where the model can be deployed.

Acknowledgements

Authors are grateful to the reviewers of this paper, due to their comments, this paper could be in the present form.

References

Adomian, G. (1990). A review of the decomposition method and some recent results for nonlinear equations. *Mathematical and Computer Modelling*, 13(7), 17–43.

Alkahtani, B. S. T., Atangana, A., and Koca, I. (2017). New nonlinear model of population growth. *Plos One*, 12(10), e0184728.

Ausloos, M. (2006). The Logistic Map and the Route to Chaos: From the Beginnings to Modern Applications. Springer Science & Business Media.

Ayata, M. and Ozkan, O. (2020). A new application of conformable laplace decomposition method for fractional newell-whitehead-segel equation. *AIMS Mathematics*, 5(6), 7402–7412.

Elgazery, N. S. (2008). Numerical solution for the falkner–skan equation. *Chaos, Solitons & Fractals*, 35(4), 738–746.

Homan, K. (2010). Athletic-ideal and thin-ideal internalization as prospective predictors of body dissatisfaction, dieting, and compulsive exercise. *Body Image*, 7(3), 240–245.

Hosseinzadeh, H., Jafari, H., and Roohani, M. (2010). Application of laplace decomposition method for solving klein-gordon equation. *World Applied Sciences Journal*, 8(7), 809–813.

Khan, H., Shah, R., Kumam, P., Baleanu, D., and Arif, M. (2020). Laplace decomposition for solving nonlinear system of fractional order partial differential equations. *Advances in Difference Equations*, 2020(1), 1–18.

Khan, Y. and Faraz, N. (2011). Application of modified Laplace decomposition method for solving boundary layer equation. *Journal of King Saud University-Science*, 23(1), 115–119.

Khuri, S. A. (2001). A laplace decomposition algorithm applied to a class of nonlinear differential equations. *Journal of Applied Mathematics*, 1(4), 141–155.

Ogunsola, A. W., Oderinu, R. A., Taiwo, M., and Owolabi, J. A. (2022). Application of laplace decomposition method to boundary value equation in a semi-infinite domain. *International Journal of Difference Equations (IJDE)*, 17(1), 75–86.

Pomeranz, S. B. (2000). The nature of mathematical modeling. by neil gershenfeld. *The American Mathematical Monthly*, 107(8), 763–766.

Ramirez-Cando, L. J., Alvarez-Mendoza, C. I., and Gutierrez-Salazar, P. (2018). Verhulst-pearl growth model versus malthusian growth model for in vitro evaluation of lead removal in wastewater by photobacterium sp. *F1000Research*, 7(491), 491.

Rani, D. and Mishra, V. (2019). Solutions of volterra integral and integro-differential equations using modified laplace adomian decomposition method. *Journal of Applied Mathematics, Statistics and Informatics*, 15(1), 5–18.

Rocha, L. S., Rocha, F. S., and Souza, T. T. (2017). Is the public sector of your country a diffusion borrower? empirical evidence from Brazil. *PloS One*, 12(10), e0185257.

Schiff, J. L. (1999). The Laplace Transform: Theory and Applications. Springer Science & Business Media.

Shah, R., Khan, H., Arif, M., and Kumam, P. (2019). Application of laplace–adomian decomposition method for the analytical solution of third-order dispersive fractional partial differential equations. *Entropy*, 21(4), 335.

Verhulst, P. F. (1838). Notice sur la loi que la population suit dans son accroissement. *Correspondence Mathematical Physics*, 10, 113–126.

Yin, X. and Zelenay, P. (2018). Kinetic models for the degradation mechanisms of PGM-free ORR catalysts. *ECS Transactions*, 85(13), 1239.

Yusufoğlu, E. (2006). Numerical solution of duffing equation by the laplace decomposition algorithm. *Applied Mathematics and Computation*, 177(2), 572–580.

7 Particle swarm optimization for wind farm optimization considering design parameters in various constraints

Vaishali Shirsath[1,a] and Prakash Burade[2,b]

[1]Poornima University, Department of computer science and Engineering, Jaipur, Rajasthan

[2]Sandip University, Department of computer science and Engineering, Nasik, India

Abstract

Wind farm related layout, it's design, along with WTG placement are much important issue as on today. Keeping the distance between the turbines, site area as well as shape are the greatest challenging assignment in front of researcher from last two decades. Using the appropriate WTG design or selecting the appropriate WTG can resolve these issues. Another set of design considerations that contribute to ensuring that the turbine fails as little as possible in order to maximize wind power generation, improve reliability, and, consequently, generate more wind power In order to avoid ambiguity when selecting the appropriate approach to solving the optimization problem, the appropriate data collection strategies for the parameters in the objective function must be aligned for the necessary simulation of the optimization problem. The square wind farm, in addition to the circle-in-circle and line-in-circle methods, has fewer solutions and may not initially attract intelligent tools. However, the rectangular shape may require the intelligent tool because it may have multiple solutions. Taking care of the particle swarm optimization (PSO) and it's trying as well as processability is most significant undertaking in this research article.

Keywords: Optimization, particle swarm optimization, wind energy, WTG

Introduction

The better method for selecting a wind farm of the right size and shape is the goal of this research. The goal is to get the most out of wind-based farm's chosen shape in-terms of annual energy production. Stanley et. al. (2019) considered fatigue loads on turbines as a constraint for the layout optimization problem in order to complete the same task. It is noted in many exploration article that breeze is one of the most reassuring wellspring of elective energy. Kusiak and Song (2009) stated that, the location of a WTG ought to be determined by the wind distribution. Wake loss is the focus of a number of wind models, most of which are based on wind turbine positions (x, y) as well as wind speed and direction (d). The greedy improvement heuristic method is used to achieve profit maximization-based WTG placement, which is based on price efficiency of the generators in addition to the power output from the wind farm referred in Aytun and Noran (2004). A set of test problems can be used to verify the heuristic's effectiveness. It has been resolved how the choppiness properties of wind

[a]vaishups2k@gmail.com, [b]prakash.burade@gmail.com

DOI: 10.1201/9781003450917-7

turbine wakes have changed after some time. Simple analytical equations based on experimental findings and numerical data from a CFD code, augmented by some hypothetical considerations, are used to suggest the choppy kinetic based energy(E) and indulgence rate given by Crespo and Hernandez (1996). In order to gain turbulence bands in the wake. Kim and Kim (2021) uses control system operation data to present NN based method for fault tolerance prediction. The selection of the data (signal points) that will serve as the neural network's input is the first step. The design of the neural network's structure is the second step. DBMS identifies the issue of optimizing wind farm architecture to position the appropriate number of WTG for maximum output power by Shirsath and Burade (2022). In any case, autonomous tasks of dispersed energy sources, like breeze, don't guarantee trustworthy power creation, for the most part because of the irradiance of the sun and the breeze's openness being likely to risk. Thus, an entirely reliable wellspring of electrical energy can be arranged by consolidating the development of wind and sunlight based energy by the team of Kumar and Khan (2021). The distribution of wind turbines at a given location is optimized to obtain the most energy for the least amount of money spent on installation.

The optimization is created using a genetic search algorithm and a wake superposition-based wind farm simulation model by Mosetti et. al. (1994). A crucial step in the design of a wind-farm is in which entails positioning the turbines within the wind farm in the best possible locations to minimize wake effects and maximize predicted power generation. The current solutions do not fully satisfy the demands of a wind farm developer Frehlich and Kelley (2010), despite the fact that the scientific community has been paying more and more attention to this problem. Wind and it's speed recorded data are initially categorized by sector of direction to accomplish this. The constant functions are mostly then used to fit the Weibull parameters to each direction sector.

The probability-density-function also known as PDF for wind direction and the turbine power characteristics are both altered into continuous functions. To take into account fetch-related variations in wind speed, the function can also be altered. When shared with wind farm cost evaluations, the levelized energy fee remains merely a function of turbine site and can therefore be employed as an objective function in an assortment of optimization methods discussed by Lackner and Elkinton (2007). Sonwane et. al. (2021) used genetic algorithm (GA) to solve the similar problem of managing both active and/or reactive power within the power transmission system: optimal allocation of DG placement. The Ministry of New and Renewable Energy Sources of the Indian Government has begun planning for the expansion of the renewable energy sector with significant modifications.

The most suitable smart power microgrids by 2022, the ministry intends to double the amount of electricity produced by renewable sources in renewable

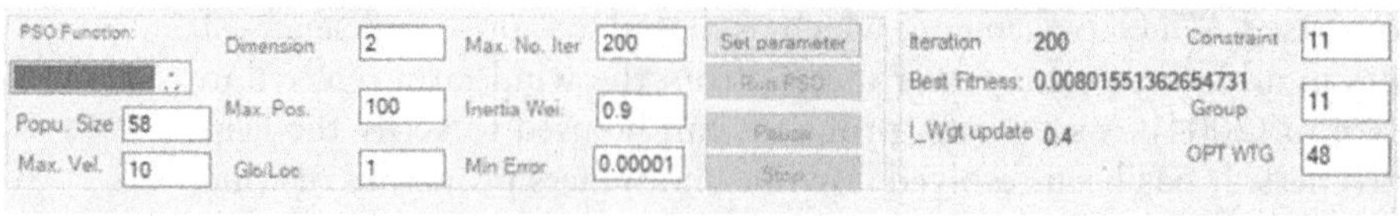

Figure 7.1 PSO Tool data insertion

electricity by Sharma and Deshmukh (2022). India will surpass other industrialized nations to become the world's leading producer of green energy if these lofty objectives are met. Renewable energy should account for 40% of India's total electricity production, according to the government. The most suitable smart power microgrids for monitoring the wind farm's generated energy are those equipped with power line communication by Rathor et. al. (2020). The wind turbine's brain is the gearbox, which is made up of multiple stages of helical/planetary gears. Performance data is gathered separately for each of these stages, and lifetime use estimation (LUE) is then calculated. From each individual LUE, the total LUE for the gearbox is then calculated. Accordingly, the flimsy part that will flop first and the disappointment mode that is causing the essential disappointment can be found. Finally, plans for the necessary corrective actions can be made. The cumulative damage and LUE are evaluated using the inverted power law damage model and Miner's rule discussed by Srinivasan and Paul Robert (2021).

A genuine offshore WF project layout optimization analysis is the subject of the existing study. In the initial step, a hereditary calculation based improvement model with constant format portrayal is made to find the best plan given the area of the WTs. In contrast to the irregular design of the horsrev1 offshore WF and the reference, this strategy's efficacy is demonstrated. Five commercial WTs are taken into consideration in the II-step to examine how different WT types affect WF goals having the fruitful results. The results showed that the best design layout is one in which large WTs are used to create WF. Include the number of W-Ts that knowingly affect both power generation and WF cost by Charhouni et. al. (2019). The authors of this research article looked at the category of wind farm shape in terms of overall profit based on farm shape, as well as the selection of wind turbines and profits associated with them. The shape category of a wind farm has an impact on the cost of the site and the optimization of the available

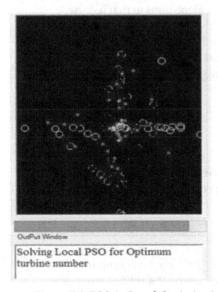

OutPut Window

Solving Local PSO for Optimum
turbine number

Figure 7.2 PSO in Local Optimization mode

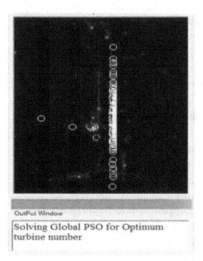

OutPut Window

Solving Global PSO for Optimum
turbine number

Figure 7.3 PSO in Global Optimization Mode

land, as land costs can often be significantly higher. Contribution of authors
here is to create four categories of farm shape as discussed here. The objective
function used here is based on the site category selection in addition to generalized constraint such as wind velocity, turbine size and type. The cost function is
operated with the particle swarm optimization (PSO) for which PSOFEED input
sources are developed and the best suitable output is suggested.

Particle swarm optimization

One of the bio-inspired algorithms is PSO which searches for the best solution in the
solution space with ease. It differs from other optimization algorithms in that it only
requires the obj-function or cost function and does not rely on the pitch or any differential objective form. Additionally, it has somehow few hyper parameters. Figure
7.1 indicates the setting of PSO parameters. Figures 7.2 and 7.3 represents the solution in local or global optimization mode. Both figures are shown here for samples.

$$V^i(t+1) = wV^i(t) + C_1r_1(pbest^i - X^i(t)) + C_2r_2(gbest - X^i(t)) \quad (1)$$

$$x^i(t+1) = x^i(t) + v_x^i(t+1) \qquad (2)$$

Before application, PSO was tested with the Rosenbrock equation and found
successful. It is noted that the PSO is mainly working on following two equations known for particle velocity and particle position.

PSOFEED equation

PSOFEED equation are required to place to resolve the objective functions
obtained herewith in the algorithms. Following are the set of PSOFEED equations as a sample to understand how the objectives are converted to PSO feed to

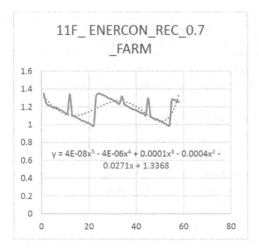

Figure 7.4 ENERCON as WTG for Rectangular Shape used in PSO Feed Table

Figure 7.5 Best fitness function as sample of farm optimization for case number 1 to 5

find the local or global best solutions for placing appropriate number of WTG and appropriate size or shape of wind farm. This table 7.2 is combinations of many functions due to different constraints and objectives available within the problem statement. Basically farm and energy are different mode of objectives in addition to both constraints. These three constraints lead to six solutions for local and global outcomes.

Result and discussion

Following are the discussion along with results are tabulated. Fifteen cases was checked and optimized using PSO tool. Table 7.2 indicates the result with

Table 7.1: PSOFEED table

Bus$_i$D	Av	Bv	Cv	Dv	Ev	Fv	Particle
1	0.000000001	-1.00E-07	0.000003	-0.00006	-0.0005	1.0245	37
2	4.00E-10	-3.00E-08	3.00E-07	-0.00001	-0.0005	1.0149	37
3	0.000000001	-1.00E-07	0.000003	-0.00005	-0.0006	1.0238	37
4	-0.000000001	0.0000001	-7.00E-06	0.0001	-0.0015	1.0531	37
5	-4.00E-10	4.00E-08	-2.00E-06	0.00002	-0.0011	1.0694	37
6	-0.000000002	0.0000003	-0.00001	0.0002	-0.0134	1.3648	49
7	0.00000003	-0.000002	0.00006	-0.0005	-0.0076	0.9872	49
8	0.00000004	-0.000003	0.00007	-0.0004	-0.012	0.9877	49
9	0.00000005	-0.000004	0.0001	-0.001	-0.0126	1.4767	49
10	0.00000007	-0.000006	0.0002	-0.001	-0.0235	1.7359	49

Table 7.2: WTG placement

Case	Farm local PSO	Energy local PSO	Both local PSO	Farm global PSO	Energy global	Both global
1	31	17	6	7	18	32
2	38	29	22	48	44	8
3	8	20	55	56	37	15
4	9	23	51	11	32	18
5	48	18	41	1	9	24
6	5	35	27	11	31	36
7	19	39	7	9	2	35
8	50	6	2	6	8	57
9	1	24	29	38	18	41
10	7	41	24	43	1	12
11	16	36	8	29	4	15
12	4	28	26	41	49	27
13	29	2	19	24	36	29
14	38	32	40	43	18	44
15	8	35	44	31	6	38

number of WTG Placement to get farm optimization or energy optimization or both cases optimization in local as well as global way.

Conclusion

This work was done for the various case studies with the right strategy for placing wind turbines in the chosen land area. According to the calculations presented in this research article, the "circle in circle" and "circle in line along with square land method" offers a novel solution that requires fewer turbine placements. With a rectangular shape, a wide range of WTG variations are possible, as seen in cases 5 through 15. With case number 13, maximum energy production is possible, but the distance between WTG and transformer station lacks internal roads and space, making it less reliable due to longer failure times.

References

Aytun Ozturk, U. and Noran, B. (2004). Heuristic methods for wind energy conversion system positioning. *Electric Power Systems Research*, 70, 179–185.

Bhatt, P. K. and Sharma, R. (2022). Optimal design analysis of multimodal hybrid AC/DC microgrids. *Mathematical Statistician and Engineering Applications*, 71(2), 138.

Charhouni, N., Sallaou, M., and Mansouri, K. (2019). Realistic wind farm design layout optimization with different wind turbines types. *International Journal of Energy and Environmental Engineering*, 10, 307–318.

Crespo, A. and Hernandez, J (1996). Turbulence characteristics in wind- turbine wakes. *Journal of Wind Engineering and Industrial Aerodynamics*, 61, 71–85.

Frehlich, R. and Kelley, N. (2010). Applications of scanning doppler lidar for the wind energy industry. *In The 90th American Meteorological Society Annual Meeting. Atlanta, GA.*

Kim, D. H. and Kim, Y. S. (2021). Failure prediction of wind turbine using neural network and operation signal. *International Journal of Recent Technology and Engineering (IJRTE)*, 10(4), 261–268. ISSN: 2277- 3878 (Online).

Kumar, A. and Khan, I. (2021). Harmonics analysis and enhancement of power quality in hybrid photovoltaic and wind power system for linear and nonlinear load using 3. *International Journal of Recent Technology and Engineering (IJRTE)*, 10(1), 296–307. ISSN: 2277-3878 (On-line).

Kusiak, A. and Song, Z (2009). Design of wind farm layout for maximum wind energy capture. *Renewable Energy*, 35, 685–694.

Lackner, M. A. and Elkinton, C. N. (2007). An analytical framework for offshore wind farm layout optimization. *Wind Engineering*, 31, 17–31.

Mosetti, G., Poloni, C., and Diviacco, D. (1994). Optimization of wind tur- bine positioning in large windfarms by means of a genetic algorithm. *Journal of Wind Engineering and Industrial Aerodynamics*, 51, 105–116.

Rathor, B., Jain, G., Jain, R., and Sonwane, P. (2020). Smart modern AC microgrid monitoring system using power line communication, 2020. *In 5th IEEE International Conference on Recent Advances and Innovations in Engineering (ICRAIE)*, (pp. 1–5) doi: 10.1109/ICRAIE51050.2020.9358274.

Sharma, P. and Deshmukh, S. (2022). A review on renewable energy in India: mission 2022. *In 2nd National Conference Recent Innovations in Science and Engineering (NC-RISE 17)*, ISSN: 2321-8169. (Vol. 5(9), pp. 32–37).

Shirsath, V. and Burade, P. (2022). Improvement of wind energy systems by optimizing turbine sizing and placement to enhance system reliability Iraqi. *Journal for Electrical and Electronic Engineering*, 18(2), 53–59.

Sonwane, P. M., Gakhar, P., Varshney, T., Garg, N. K., Jayaswal, K., and Jain, G. (2021). Optimal allocation of distributed generator placement: an optimal approach to enhance the reliability of micro-grid. *In 6th IEEE International Conference on Recent Advances and Innovations in Engineering (ICRAIE)*, 2021.

Srinivasan, R. and Paul Robert, T. (2021). Remaining useful life prediction on wind turbine gearbox. *International Journal of Recent Technology and Engineering (IJRTE)*, 9(5), 57–65. ISSN: 2277-3878 (Online).

Stanley, A. P. J., King, J., and Ning, A. (2019). Wind farm layout optimization with loads considerations. *Journal of Physics. In NAWEA WindTech 2019 IOP Publishing.*

8 Non-conventionally engineered nanoparticles as potential remedial towards eco-sustainability

Anju Yadav[1,a], Shruti Sharma[2,b], Rekha Nair[3,c] and Alka Sharma[1,d]

[1]Department of Chemistry, University of Rajasthan, Jaipur (Rajasthan), India

[2]DIC, CCT, University of Rajasthan, Jaipur (Rajasthan) India

[3]Poornima College of Engineering, Jaipur (Rajasthan) India

Abstract

Industrial wastewater effluents are the well-known biggest source of hazardous, non-biodegradable organic pollutants inevitably contaminates water bodies, soil, and atmosphere. Nanotechnology plays a pivotal role as remedial strategies to severe environmental threats and a wide variety of specifically designed nanomaterials, with respect to dimension, morphology, chemical composition and state, can be produced with enhanced properties as compared to the traditional bulk material. A broad spectrum of nanomaterials has been synthesized via conventional routes, however, this may be a threat to the ecosystem on account of harmful synthetic chemicals used during the course of synthesis. In view of environment legislations, the nanomaterials are produced via non-conventional approaches using natural resources. Indigenous plant, *Holoptelea integrifolia*, a significant pollen allergen that sensitizes 10% of the atopic population in India, yet richly constituted with bioactive phytochemicals to act as reductant, stabilizer and capping agents, was used to fabricate quite a few nanoparticles having efficacy to act as catalyst. In this succession, copper nanoparticles (CuNPs) were successfully fabricated non-conventionally employing extract of *Holoptelea integrifolia* fruits. Fabricated CuNPs was probed for their photocatalytic efficacy to degrade and mineralize toxic, non-biodegradable and most aquatic-pollutant methylene blue (MB) dye. Promising degradation efficiency (96%) was observed. The kinetics investigation resulted a pseudo-first-order reaction with 0.016 min^{-1} rate constant. Impact of variable parameters, *viz.*, photocatalyst dosage, dye concentration, sunlight-exposure time, on photocatalytic efficiency of CuNPs was also investigated. Thereby, concluded that the non-conventionally fabricated *green* CuNPs have significantly high potency to eradicate non-biodegradable, hazardous and most aquatic-pollutant dye, MB, and thus, can act as proficient remediation for sustainable environment.

Keywords: Copper nanoparticles, industrial wastewater effluent, methylene blue, photocatalytic degradation, potable water

Introduction

Nanostructure materials have been studied very vividly in the recent times due to their distinctive characteristics including large surface area as compared with

[a]mehtaanju77@gmail.com, [b]shruti.chaturvedi15@gmail.com, [c]rekhanair@poornima.org, [d]sharma_alka21@yahoo.com

DOI: 10.1201/9781003450917-8

the bulk materials. Metal nanomaterials, in particular, are extensively researched owing to their unusual optical, catalytic, and electromagnetic capabilities. In numerous industries, including medicine, therapeutics, ecology, catalysis, sensing, and optoelectronics, metal nanoparticles have been widely used (Biao et. al., 2017; Liang et. al., 2017; Nagajyothi et. al., 2017; Tang et. al., 2018).

Many nanomaterials have been created via traditional methods, however because these methods employ substances that could be toxic to the environment. Therefore, ecologically friendly, cost-effective, non-toxic, non-conventional (green) methods have become increasingly popular in recent years for producing nanomaterials (Abbas and Fairouz, 2022; Ahmed et. al., 2019; Ajmal et. al., 2016; Anju et. al., 2020; 2021; 2022; Fatima and Wahid, 2022; Issaabadi et. al., 2017; Jayarambabu et. al., 2020; John et. al., 2021; Kandasamy et. al., 2020; Kayalvizhi et. al., 2020; Letchumanan et. al., 2021; Mousa et. al., 2022; Nagajyothi et. al., 2017; 2020; Nasrollahzadeh and Mohammad, 2015; Nguyen et. al., 2021; Safajou et. al., 2021; Singh et. al., 2019; Shubhashree et. al., 2022; Sharma et. al., 2022; Sriramulu et. al., 2020; Vasantharaj et. al., 2019; 2019a).

Considering its capabilities and prospective uses in a number of industries, copper nanoparticles (CuNPs) have garnered the most attention among the many other metal NPs, *viz.*, catalysis, optics, solar cells, energy conversion, biomedical, antibacterial activities, and dye degradation. CuNPs have also been successfully synthesized via non-conventional routes (Abbas and Fairouz, 2022; Ahmed et. al., 2019; Chandrasekaran, 2013; Fatima and Wahid, 2022; Issaabadi et. al., 2017; Jayarambabu et. al., 2020; John et. al., 2021; Kayalvizhi et. al., 2020; Letchumanan et. al., 2021; Nagajyothi et. al., 2017; Nasrollahzadeh and Mohammad, 2015; Safajou et. al., 2021; Singh et. al., 2019; Sriramulu et. al., 2020; Vasantharaj et. al., 2019). The bio-active phytochemicals, *viz.*, flavonoids, terpenoids, polyphenols, etc. perform actions to reduce, cap and stabilize, thereby, enable to fabricate nanomaterials (Anju et. al., 2020; 2021; 2022; Nasrollahzadeh and Mohammad, 2015; Sharma et. al., 2022).

Science and technology advancements has steered a better living of humanity and strengthened the economy of any country. But, during this techno savvy and modernization process, a significant amount of organic pollutants and other toxic chemicals gets generated and released, thereby disturbing the environment. In addition to lot many other industries, textile, paint, paper, cosmetics, drug and pharma industries have been known as the major contributors of several hazardous pollutants. Many dyes have been abundantly used in such industries and with the release of their effluents, it contaminates soil, water-bodies, and the environment. As a result, this impacts on aquatic bodies, vegetation and living beings as well. Such devastation has been minimized through elimination/degradation of the hazardous organic dyes by employing various conventional techniques, *viz.*, adsorption (Aramesh et. al., 2021; Baloo et. al., 2021), ultrafiltration (Ding et. al., 2021), coagulation (Birjandi et. al., 2016), membrane separation (Yu et. al., 2017; Zhang et. al., 2018; Moradihamedani, 2022), intermittent aeration strategy (Oliveira et. al., 2020; Wu et. al., 2016), and photocatalytic degradation (Ajmal et. al., 2016; Liang et. al., 2017; Loghambal et. al., 2018; Ma et. al., 2020; Mousa et. al., 2022; Nagajyothi et. al., 2020; Nguyen et. al., 2021;

Safajou et. al., 2021; Salama, 2017; Saini et. al., 2020; Sathiyavimal et. al., 2020; Singh et. al., 2019; Vasantharaj et. al., 2019; 2019*a*; Vinayagam et. al., 2022). Among all these approaches, photocatalytic degradation is known to be the most effective, and cost-effective to degrade dye-pollutants and prevent environmental contamination. In the present study, methylene blue (MB), one of the most carcinogenic, poisonous, not recyclable, and water-polluting dyes, was successfully degraded using CuNPs.

Materials and methods

Sigma-Aldrich make AnalR grade copper sulfate pentahydrate (CuSO$_4$.5H$_2$O) and methylene blue (C$_{16}$H$_{18}$N$_3$SCl) were procured for their high purity. The fresh fruits of *Holoptelea integrifolia* were congregated from the University of Rajasthan, Jaipur campus. All solutions were prepared using doubly distilled water and experimentation was carried out at room temperature.

Holoptelea integrifolia fruit-extract-mediated copper nanoparticles

Holotelea Integrifolia fresh fruit (Figure 8.1) was gathered from the university's campus, cleaned using tap water, and shade dried for 14–15 days at room temperature. Using a mixer grinder, dried fruits were finely powdered to be used for extraction process.

Soxhlet extractor was used to prepare aqueous extract as per the standard procedure (Anju et. al., 2020; 2021; 2022) and stored at room temperature in an airtight container for experimentation. Copper nanoparticles (CuNPs) were fabricated as per the reported procedure (Anju et. al., 2020) via non-conventional, cost-effective, and eco-friendly greener route, avoiding use of any harmful chemicals. The photocatalytic activity of these non-conventionally produced CuNPs were examined to destroy the most prevalent aquatic pollutant, methylene blue (MB) dye, utilizing fruit extract from *Holoptelea integrifolia*.

Preparation of stock solution of methylene blue dye

To make its stock solution (1000 mg/L), 1g of commercially available methylene blue dye (AR grade) was diluted in 1000 mL in doubly distilled water.

Figure 8.1 Holoptelea integrifolia fruits

Degradation of aquatic-pollutant (MB) dye by copper nanoparticles

After successful fabrication of CuNPs (employing *Holoptelea integrifolia* fruit-aqueous extract), the photocatalytic efficacy of these CuNPs were investigated towards the degradation of the most aquatic-pollutant dye (MB) under exposure of sunlight. To determine the optimum values of dye concentration, catalyst dose, and solar irradiation period. The dye concentration, catalytic dose, and irradiation period were three variables that were taken into consideration, as per the standard reported procedure (Ahmed et. al., 2019; Anju et. al., 2022; Kayalvizhi et. al., 2020; Liang et. al., 2017; Mousa et. al., 2022; Nagajyothi et. al., 2020; Nguyen et. al., 2021; Safajou et. al., 2021; Salama, 2017; Saini et. al., 2020; Sathiyavimal et. al., 2020; Singh et. al., 2019; Vasantharaj et. al., 2019; 2019a; Vinayagam et. al., 2022).

Using MB dye stock solution, a variety of MB dye test solutions (5–25 ppm) were created using water as the solvent. The % degradation was examined for the various dye-concentrations in the range (5–25 ppm) so as to acquire the optimum concentration of dye (15 ppm). Similarly, a varied range (5–25 mg) of photocatalyst (CuNPs) dosages was examined to obtain its optimum dose (20 mg); and photocatalytic degradation of MB dye by the fabricated nano catalyst was investigated under a range of sunlight-exposure time (15 min to 150 min i.e., upto 2.5 hrs) at fixed time intervals @15 min, to acquire the optimum sunlight-exposure time.

After acquiring the optimum values, 20 mg of fabricated CuNPs was added to 15 ppm of MB dye suspension. The adsorption-desorption equilibrium, was uphold by stirring the reaction mixture thoroughly for 30 min in the dark using an incubator shaker prior to the initiation of photocatalytic process by solar-irradiation f. After a stipulated time period (15 minutes), a preset volume of the reaction mixture (2–3 mL) was taken out, filtered using 0.22-m Nylon syringe filter, and centrifuged for five minutes at 5000 rpm. This reaction mixture (sample) was investigated for the absorbance recording using UV-vis spectrophotometer (Thermo Fisher Scientific MULTISKAN GO). The aforementioned process was repeated every 15 minutes for a total of 2.5 hours. To evaluate the degraded amount of MB dye, the equation below was used.

$$\text{Percentage degradation (\%)} = \left(\frac{C_0 - C_t}{C_0} \right)$$

where, correspondingly, C_0 represents the dye concentration before degradation and C_t represents the dye concentration at various exposure times "t" in the aqueous phase.

Results and discussion

Ability of non-conventionally manufactured CuNPs (Anju et. al., 2020) to breakdown the non-biodegradable and most hazardous aquatic dye, MB, was examined. This was done under exposure to direct sunshine. The investigatory data showed very encouraging % degradation efficacy of the photocatalyst (CuNPs) toward MB dye degradation.

Photocatalytic efficacy of non-conventionally fabricated copper nanoparticles

Photocatalytic efficacy of non-conventionally fabricated CuNPs was investigated to degrade most aquatic-pollutant dye (MB). The dye degradation takes time, and it was monitored by measuring changes in the reaction mixture's absorbance using UV-vis spectrophotometer. MB dye degradation using photocatalysts (CuNPs) was investigated for the effects of three variables, the dye concentration, catalyst dose, and sunlight exposed period, so as to optimize dye concentration (ppm), catalyst dose (mg), and solar irradiation time (h). UV spectrum's absorption peak of MB occurs at 664 nm and in presence of photocatalyst (CuNPs), the absorption peak of MB declined steadily with a minimum at the sunlight-exposure period 2.5 hrs. The % dye-degradation efficacy of 20 mg of CuNPs was observed maximum (96%) for 15 ppm of MB dye.

Effect of variation in catalyst load

To acquire the optimum dose of catalyst (CuNPs) (20 mg), a varied range of catalyst's dose (5–25 mg) was investigated for its photocatalytic efficacy keeping the dye concentration and other parameters constant. The effect of catalyst (CuNPs) load on its % degradation efficacy towards MB dye has been illustrated in Figure 8.2. The % dye-degradation rises with increase in the catalyst dose and is found to be maximum (96%) at 20 mg dose of the catalyst. With further increase in the catalyst dose (25 mg), the % degradation of dye was found to decline (Figure 8.2). Consequently, the optimum dose of the catalyst (CuNPs) was acquired as 20 mg with its maximum % degradation efficacy.

The trend observed (Figure 8.2) for catalyst-dose vs % degradation may be elucidated on connecting with the available exposed surface area vs reaction rate. With an initial increase in catalyst dose, the catalyst-surface area increased, thereby enhance in rate of reaction and the % degradation. However, further increase in the catalyst-dose (beyond its optimum value) lead to increase in the thickness of the layer at the base-end of reaction pot entirely roofed by the catalyst, instead of surge in exposed surface area. Thereby, cause retardation in the reaction rate beyond optimum dose of catalyst.

Effect of variation in methylene blue dye concentration

Using water as solvent, varied concentrations (5–25 ppm) of MB dye solutions were prepared from its stock solution. Keeping all other parameters identical, each test solution consisting of a fixed (20 mg) catalyst dose were examined to detect the dye concentration-effect on the % degradation and consequently, an optimum concentration of MB dye for maximum % degradation (Figure 8.3) was obtained. Initially, the % degradation was found to decrease up to 10 ppm of MB dye concentration, later it increases and gets maximum (95%) at 15 ppm of MB dye. On increasing the MB dye concentration, a sharp decline in the

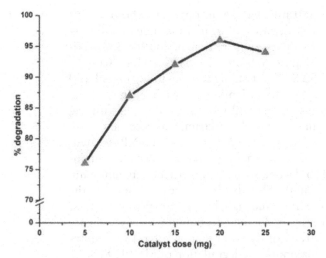

Figure 8.2 Effect of catalyst (CuNPs) load on its % degradation efficacy for MB dye.

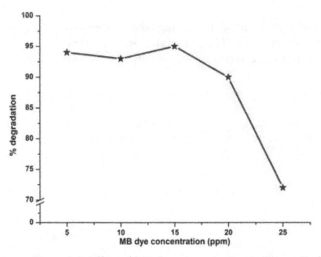

Figure 8.3 Effect of MB dye concentration (ppm) on % degradation

% degradation was observed with lowest % degradation (75%) at 25 ppm of MB dye. Thereby, the optimum MB dye concentration was acquired as 15 ppm which resulted in the highest % degradation (Figure 8.3). This pattern may be explained by the relationship between reaction rate and the number of dye molecules., but further increase in concentration resulted decrease in reaction rate due to increase in collisions between dye molecules and decrease in collisions between dye and OH⁻ radicals.

Effect of variation in sunlight-irradiation time

Variations in sunlight exposure time and its effect on the catalysts' ability to degrade MB dye by photo catalyst was investigated. Test solution with fixed concentration (15 ppm) of MB dye (80 mL) having an optimum photocatalyst dose (20 mg) were investigated over a wide range of irradiation time period (15–150 min), and an optimum value of sunlight-exposure time period (hr) for maximum % degradation was acquired (Figure 8.4a and b). The catalyst (CuNPs) degraded MB dye to its maximum (92.88%) at sunlight-irradiation time of 150 min. After a predetermined amount of time (15 minutes), took 23 mL of the reaction mixture, filtered using Nylon syringe (0.22 μm) filter. After centrifuging for 5 min @ 5000 rpm the supernatant's UV-visible spectrum was recorded.

Figure 8.4(a) displays the UV-Vis spectra of the MB dye and the NPs-treated-dye (test) solutions. The absorption peak of reaction mixture solution containing MB dye and CuNPs was found to decline steadily and reached its minimum at maximum exposure period (150 min). The absorption peak shrinks as the exposure duration increases and after two and half hours of exposure, a faint absorption peak was observed, thereby endorsing degradation of dye. Figure 8.4(b) depicted clearly the observed trend of dye degradation after 150 minutes of sunlight irradiation and attaining maximum % degradation of MB (92.88%).

Dye degradation kinetics

The rate of photocatalytic processes was calculated using the Langmuir-Hinshelwood kinetic model (Khezrianjoo et. al., 2016; Loghambal et. al., 2018). The rate of reaction was evaluated using the expression:

$$r = \frac{dC}{dt} = k \ KC \ / 1 + KC$$

Reaction rate constant, k, for pseudo-first-order was evaluated by the expression given as:

$$k = \ln \frac{C}{C_0} \ / \ t$$

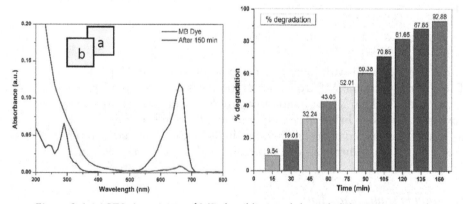

Figure 8.4 (a) UV-vis spectras of MB dye (blue) and degraded dye solutions (after 150 min sunlight-exposure) (red); and (b) % degradation of MB dye vs. exposure time @fixed intervals of 15 min till 150 min

where C_0 is the concentration of dye at the beginning of the process, C is the concentration of dye at the various irradiation durations (t), and k is the apparent pseudo-first-order rate constant (min^{-1}).

A plot between ln (C/C_0) and time period exposed to sunlight (t) evidently show linearity (Figure 8.5).

The kinetics study of deterioration of MB dye photo catalytically by CuNPs, i.e., pseudo-first order reaction, is explicitly supported by the excellent linear curve of ln C/C_0 with time produced by the kinetic calculations.

Langmuir–Hinshelwood kinetic model was found to be fitted for the photocatalytic destruction of MB dye, with its rate constant evaluated as 0.016 min^{-1} and the regression coefficient $R^2 = 0.9402$. Therefore, CuNPs has been observed to be a stable and effective *green* catalyst for the breakdown of the most aquatic-pollutant dye (MB) irradiated in sunlight.

Conceivable mechanism of photocatalytic activity of copper nanoparticles

Among many nano catalysts (metal/metal oxide), CuNPs have established to be one of the most promising nano catalysts for the degradation of the most aquatic-pollutant dye (MB). The conceivable photocatalytic mechanism of CuNPs as catalyst to degrade most aquatic-pollutant MB dye has been sketched (Figure 8.6).

On irradiation with the sunlight, CuNPs exhibited excitation, generating duos of holes (h$^+$) and free electrons (e$^-$). The free electrons (e$^-$) transforms the available adsorbed molecular oxygen (O$_2$) into superoxide radicals (O$_2^{-\bullet}$), whereas the holes (h$^+$) interacted with water (H$_2$O) thus forming H$^+$ and OH$^\bullet$ ions.

The available newly formed superoxide (O$_2^{-\bullet}$) radicals react with the water molecules to produce H$_2$O$_2$ and O$_2$, and successively, on irradiation with

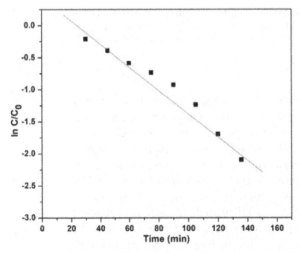

Figure 8.5 Plot of linear relationship between ln(C/C0) vs. sunlight exposed time (min) for the catalytic decay of MB dye

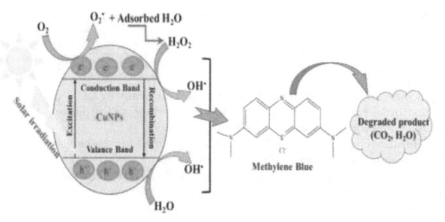

Figure 8.6 Proposed mechanism of CuNPs aided photocatalytic action to degrade MB dye

sunlight, these gets transformed in to OH• radicals (Ajmal et. al., 2016). Thereby, it can be explicitly endorsed that the pairs of free electrons-holes play a significant role to generate potent oxidant (OH• radicals) which degrade or destroy MB dye molecules to low weighed non-toxic organic moiety(s), mineral ions, water and carbon dioxide molecules, etc. By probing the deteriorated residues, it was discovered that the MB dye molecules were destroyed and mineralized during the degradation process rather than transformed into new contaminants or forms. Credibly, the non-conventionally fabricated CuNPs, using *Holoptelea integrifolia* fruit extract, has proved to possess potential photocatalytic activity towards degradation of the carcinogenic, non-biodegradable, most aquatic-toxicant pollutant (MB) dye. Moreover, the industry-released dye-effluent wastewater could be very effectively converted into pollutant-free potable water which can be made accessible for agriculture purposes. Thus, this can be a good doable solution for the problems related to the water-scarcity zones as well.

Conclusions

Non-conventionally fabricated copper nanoparticles (CuNPs), using *Holoptelea integrifolia* fruit extract, proved to be a potential catalyst with significantly high degrading efficacy (96%) toward methylene blue (MB) dye under sunlight-exposure. Only 20 mg of fabricated *green* CuNPs as catalyst was sufficient enough to degrade 15 ppm MB dye from an aqueous medium on a solar-exposure period of 2.5 hr. The kinetic study illustrated that the photocatalytic degradation reaction by CuNPs towards MB dye molecules follow pseudo-first-order, thereby endorsing the catalytic potency of non-conventionally fabricated copper nanoparticles.

Jaipur, The World Heritage-Pink City, has many textile industries which might have implanted water-treatment plants for pollutant free wastewater disposal. However, there are lots of small scale tie and dye units which are unable to implant such water-treatment units (due to financial constraints), and dispose waste water effluents consisting a large number of dye molecules/toxic organic moieties to either soil or public-drainage systems thereby, contaminating the soil (agriculture lands), running water streams and underground water bodies. Consequently, if not treated timely over a period of time, could be lethal to humans and living beings. On a conclusive note, the green CuNPs were found to have significantly high photocatalytic proficiency and it may be effectively employed to degrade the hazardous, non-biodegradable, carcinogenic and most aquatic-pollutant (MB) dye. To make the industrial wastewater effluent potable for agriculture purposes, conserving water specifically in the water-scarcity zones and for environmental remediation as well, such non-conventionally fabricated (green) CuNPs must be adopted and also practiced at larger (industrial) levels. Moreover, this non-conventional approach is an innovative, sustainable, less-costlier, easy doable and highly success-oriented eco-remedial solution to recycle and conserve water.

Acknowledgment

Anju appreciates the financial assistance from CSIR in order to complete the thesis-related research. The heads of the departments of chemistry and zoology at the University of Rajasthan in Jaipur are sincerely thanked for providing the facilities for research and for granting access to the instruments required for this study, respectively.

Ethics approval

This is original work and not published anywhere. In the research study, neither humans nor animals are used as subjects in investigations.

Consent to participate

Not applicable

Consent to publish

This is original work.

Author(s) contributions

The research concept, experiments, data analysis, data interpretation, results, writing, and submission for publication of this research report were all assisted by the authors Alka Sharma, Anju, and Shruti Sharma. Rekha Nair helped with certain aspects of the writing, editing, and preliminary experimenting for this study publication.

Funding

Not applicable

Competing interest

There are no conflicting interests listed by the authors.

Availability of data and materials

Not applicable

References

Abbas, A. H. and Fairouz, N. Y. (2022). Characterization, biosynthesis of copper nanoparticles using ginger roots extract and investigation of its antibacterial activity. *Materials Today: Proceedings*, 61(3), 908–913. http://doi.org/10.1016/j.matpr.2021.09.551

Ahmed, A., Usman, M., Liu, Q.-Y., Shen, Y.-Q., Yua, B., and Cong, H.-L. (2019). Plant mediated synthesis of copper nanoparticles by using *Camelia sinensis* leaves extract and their applications in dye degradation. *Ferroelectrics*, 549(1), 61–69. http://doi.org/10.1080/00150193.2019.1592544

Ajmal, A., Majeed, I., Malik, R. N., Iqbal, M., Arif Nadeem, M., Hussain, I., Yousaf, S., Zeshan, Mustafa, G., Zafar, M. I., and Amtiaz Nadeem, M. (2016). Photocatalytic degradation of textile dyes on Cu$_2$O-CuO/TiO$_2$ anatase powders. *Journal of Environmental Chemical Engineering*, 4(2), 2138–2146. https://doi.org/10.1016/j.jece.2016.03.041

Anju, Sharma, S., Dhanetia, H. R., and Sharma, A. (2020). Green synthesis of copper nanoparticles using *Holoptelea integrifolia* fruit extract. *RASĀYAN Journal of Chemistry*, 13(4), 2664–2671. http://dx.doi.org/10.31788/RJC.2020.1346306.

Anju, Sharma, S., Dhanetia, H. R., and Sharma, A. (2021). Green synthesis and characterization of cerium oxide nanoparticles using *Euphorbia hirta* leaf extract. *Journal of International Academy of Physical Sciences*, 25(4), 579–590. ISSN: 0974-9373.

Anju, Sharma, S., Dhanetia, H. R. and Sharma, A. (2020). Green synthesis of copper nanoparticles using *Holoptelea integrifolia* fruit extract. RASAYAN J Chem 13(4):2664–2671. http://dx.doi.org/10.31788/RJC.2020.1346306

Aramesh, N., Bagheri, A. R., and Bilal, M. (2021). Chitosan-based hybrid materials for adsorptive removal of dyes and underlying interaction mechanisms. *International Journal of Biological Macromolecules*, 183, 399–422. http://doi.org/10.1016/j.ijbiomac.2021.04.158

Baloo, L., Isa, M. H., Sapari, N. B., Jagaba, A. H., Wei, L. J., Yavari, S., Razali, R., and Vasu, R. (2021). Adsorptive removal of methylene blue and acid orange 10 dyes from aqueous solutions using oil palm wastes-derived activated carbons. *Alexandria Engineering Journal*, 60(6), 5611–5629. http://doi.org/10.1016/j.aej.2021.04.044

Biao, L., Tan, S., Wang, Y., Guo, X., Fu, Y., Xu, F., Zu, Y., and Liu, Z. (2017). Synthesis, characterization and antibacterial study on the chitosan-functionalized Ag nanoparticles. *Materials Science and Engineering: C*, 76, 73–80. http://doi.org/10.1016/j.msec.2017.02.154

Birjandi, N., Younesi, H., and Bahramifar, N. (2016). Treatment of wastewater effluents from paper recycling plants by coagulation process and optimization of treatment conditions with response surface methodology. *Applied Water Science*, 6, 339–348. http://doi.org/10.1007/s13201-014-0231-5

Chandrasekaran, S. (2013). Novel single-step synthesis, high efficiency and cost-effective photovoltaic applications of oxidized copper nanoparticles. *Solar Energy Materials & Solar Cells*, 109, 220–226. https://doi.org/10.1016/j.solmat.2012.11.003

Ding, J., Sarrigani, G. V., Qu, J., Ebrahimi, A., Zhong, X., Hou, W.-C., Cairney, J. M., Huang, J., Wiley, D. E., and Wang, D. K. (2021). Designing Co_3O_4/silica catalysts and intensified ultrafiltration membrane-catalysis process for wastewater treatment. *Chemical Engineering Journal*, 419, 129465. http://doi.org/10.1016/j. cej.2021.129465

Fatima, F. and Wahid, I. (2022). Eco-friendly synthesis of silver and copper nanoparticles by *Shizophyllum commune* fungus and its biomedical applications. *International Journal of Environmental Science and Technology*, 19, 7915–7926. http://doi. org/10.1007/s13762-021-03517-6

Issaabadi, Z., Nasrollahzadeh, M., and Sajadi, S. M. (2017). Green synthesis of copper nanoparticles supported on bentonite and investigation of its catalytic activity. *Journal of Cleaner Production*, 142(4), 3584–3591. http://doi.org/j. jclepro.2016.10.109

Jayarambabu, N., Akshaykranth, A., Rao, T. V., Rao, K. V., and Kumar, R. R. (2020). Green synthesis of Cu nanoparticles using *Curcuma long* extract and their application in antimicrobial activity. *Materials Letters*, 259, 126813. http://doi.org/10.1016/j. matlet.2019.126813

John, M. S., Nagoth, J. A., Zannotti, M., Giovannetti, R., Mancini, A., Ramasamy, K. P., Miceli, C., and Pucciarelli, S. (2021). Biogenic synthesis of copper nanoparticles using bacterial stains isolated from an antarctic consortium associated to a psychrophilic marine ciliate: characterization and potential application as antimicrobial agents. *Marine Drugs*, 19(5), 263. http://doi.org/10.3390/md19050263

Kandasamy, S., Chinnappan, S., Thangaswamy, S., and Balakrishnan, S. (2020). Facile approach for phytosynthesis of gold nanoparticles from *Corallocarbus epigaeus rhizome* extract and their biological assessment. *Materials Research Express*, 6(12), 1250c1. http://doi.org/10.1088/2053-1591/ab608f

Kayalvizhi, S., Sengottaiyan, A., Selvankumar, T., Senthilkumar, T., Sudhakar, C., and Selvam, K. (2020). Eco-friendly cost-effective approach for synthesis of copper oxide nanoparticles for enhanced photocatalytic performance. *Optik*, 202, 163507. https:// doi.org/10.1016/j.ijleo.2019.163507

Khezrianjoo S, Revanasiddappa HD (2016) Effect of operational parameters and kinetic study on the photocatalytic degradation of *m*-cresol purple using irradiated ZnO in Aqueous medium. *Water Qual Res J*, 51(1):69–78. http://doi.org/10.2166/ wqrjc.2015.028

Letchumanan, D., Sok, S. P. M., Ibrahim, S., Nagoor, N. H., and Arshad, N. M. (2021). Plant-based biosynthesis of copper/copper oxide nanoparticles: an update on their applications in biomedicine, mechanisms and toxicity. *Biomolecules*, 11, 1–27. http:// doi.org/10.3390/biom11040564

Liang, Y., Chen, Z., Yao, W., Wang, P., Yu, S., and Wang, X. (2017). Decorating of Ag and CuO on Cu nanoparticles for enhanced high catalytic activity to the degradation of organic pollutants. *Langmuir*, 33(31), 7606–7614. http://doi.org/10.1021/ acs.langmuir.7b01540

Loghambal, S., Catherine, A. A., and Subash, S. V. (2018). Analysis of langmuir-hinshelwood kinetics model for photocatalytic degradation of aqueous direct blue 71 through analytical expression. *International Journal of Mathematics and its Applications*, 6, 903–913. http://ijmaa.in/v6n1-e/903-913

Ma, X., Dang, R., Liu, J., Yang, F., Li, H., Zhang, Y., and Luo, J. (2020). Facile synthesis and characterization of spinel $NiFe_2O_4$ nanoparticles and studies of their photocatalytic for oxidation of alcohols. *Science of Advanced Materials*, 12, 357–365. http:// doi.org/10.1166/sam.2020.3549

Moradihamedani, P. (2022). Recent advances in dye removal from wastewater by membrane technology: a review. *Polymer Bulletin*, 79, 2603–2631. http://doi.org/10.1007/s00289-021-03603-2

Mousa, S. A., Shalan, A. E., Hassan, H. H., Ebnawaled, A. A., and Khairy, S. A. (2022). Enhanced the photocatalytic degradation of titanium dioxide nanoparticles synthesized by different plant extracts for wastewater treatment. *Journal of Molecular Structure*, 1250, 131912. http://doi.org/10.1016/j.molstruc.2021.131912

Nagajyothi, P. C., Muthuraman, P., Sreekanth, T. V. M., Kim, D. H., and Shim, J. (2017). Green synthesis: in vitro anticancer activity of copper oxide nanoparticles against human cervical carcinoma cells. *Arabian Journal of Chemistry*, 10, 215–225. http://doi.org/10.1016/j. arabjc.2016.01.011

Nagajyothi, P. C., Prabhakar Vattikuti, S. V., Devarayapalli, K. C., Yoo, K., Shim, J., and Sreekanth, T. V. M. (2020). Green synthesis: photocatalytic degradation of textile dyes using metal and metal oxide nanoparticles-latest trends and advancements. *Critical Reviews in Environmental Science and Technology*, 50, 2617–2723. https://doi.org/10.1080/10643389.2019.1705103

Nasrollahzadeh, M. and Mohammad, S. (2015). Green synthesis of copper nanoparticles using *Ginkgo biloba* L. leaf extract and their catalytic activity for the huisgen [3+2] cycloaddition of azides and alkynes at room temperature. *Journal of Colloid and Interface Science*, 457, 141–147. http://doi.org/10.1016/j.jcis.2015.07.004

Nguyen, D. T. C., Le, H. T., Nguyen, T. T., Bach, L. G., Nguyen, T. D., and Tran, T. V. (2021). Multifunctional ZnO nanoparticles bio-fabricated from *Canna indica* L. flowers for seed germination, adsorption, and photocatalytic degradation of organic dyes. Journal of Hazardous Materials, 420, 1–15. http://doi.org/10.1016/j.jhazmat.2021.126586

Oliveira, J. M. S., de Lima e Silva, M., Issa, C. G., Corbi, J. J., Damianovic, M. H. R. Z., and Foresti, E. (2020). Intermittent aeration strategy for azo dye biodegradation: a suitable alternative to conventional biological treatments? *Journal of Hazardous Materials*, 385, 1–38. http://doi.org/10.1016/j.jhazmat.2019.121558

Safajou, H., Ghanbari, M., Amiri, O., Khojasteh, H., Namvar, F., Zinatloo-Ajabshir, S., and Salavati-Niasari, M. (2021). Green synthesis and characterization of RGO/Cu nanocomposites as photocatalytic degradation of organic pollutants in wastewater. *International Journal of Hydrogen Energy*, 46, 20534–20546. http://doi.org/10.1016/j.ijhydene.2021.03.175

Saini, M., Brijnandan, D., and Ahmad, U. (2020). $VO_2(M)@CeO_2$ core-shell nanospheres for thermochromic smart windows and photocatalytic applications. *Ceramics International*, 46, 986–995. http://doi.org/10.1016/j.ceramint.2019.09.062

Salama, A. (2017). New sustainable hybrid material as adsorbent for dye removal from aqueous solutions. *Journal of Colloid and Interface Science*, 487, 348–353. http://doi.org/10.1016/j.jcis.2016.10.034

Sathiyavimal, S., Vasantharaj, S., Kaliannan, T., and Pugazhendhi, A. (2020). Eco biocompatibility of chitosan-coated biosynthesized copper oxide nanocomposite for enhanced industrial (Azo) dye removal from aqueous solution and antibacterial properties. *Carbohydrate Polymers*, 241, 1–11. https://doi.org/10.1016/j.carbpol.2020.116243

Sharma, S., Anju, and Sharma, A. (2022). Non-conventional fabrication of Ag@ZnO nanocomposite core-shell using extract of weed (*Euphorbia hirta*) leaves. *Journal of International Academy of Physical Sciences*, 26(3), 307–319. ISSN: 0974-9373.

Shubhashree, K. R., Reddy, R., Gangula, K. A., Nagananda, G. S., Badiya, P. K., Ramamurthy, S. S., Aramwit, P., and Reddy, N. (2022). Green synthesis of copper nanoparticles using aqueous extracts from *Hyptis suaveolens* (L.). *Materials Chemistry and Physics*, 280, 125795. http://doi.org/10.1016/j.matchemphys.2022.125795

Singh, J., Kumar, V., Kim, K. H., and Rawat, M. (2019). Biogenic synthesis of copper oxide nanoparticles using plant extract and its prodigious potential for photocatalytic degradation of dyes. *Environmental Research*, 177, 108569. http://doi.org/j.envres.2019.108569

Sriramulu, M., Shanmugam, S., Ponnusamy, V. K. (2020). *Agaricus bisporus* mediated biosynthesis of copper nanoparticles and its biological effects: an in-vitro study. *Colloid and Interface Science Communications*, 35, 1–8. https://doi.org/10.1016/j.colcom.2020.100254

Tang, Q., Yu, B., Gao, L., Cong, H., Song, N., and Lu, C. (2018). Stimuli responsive nanoparticles for controlled anti-cancer drug release. *Current Medicinal Chemistry*, 25, 1837–1866. http://doi.org/10.2174/0929867325666180111095913

Vasantharaj, S., Sathiyavimal, S., Saravanan, M., Senthilkumar, P., Gnanasekaran, K., Shanmugavel, M., Manikandan, E., and Pugazhendhi, A. (2019). Synthesis of eco-friendly copper oxide nanoparticles for fabrication over textile fabrics: Characterization of antibacterial activity and dye degradation potential. *Journal of Photochemistry and Photobiology B: Biology*, 191, 143–149. https://doi.org/10.1016/j.jphotobiol.2018.12.026.

Vasantharaj, S., Sathiyavimal, S., Senthilkumar, P., LewisOscar, F., and Pugazhendhi, A. (2019*a*). Biosynthesis of iron oxide nanoparticles using leaf extract of *Ruellia tuberosa*: antimicrobial properties and their applications in photocatalytic degradation. *Journal of Photochemistry and Photobiology B: Biology*, 192, 74–82. https://doi.org/10.1016/j.jphotobiol.2018.12.025

Vinayagam, R., Singhania, B., Murugesan, G., Kumar, P. S., Bhole, R., Narasimhan, M. K., Varadavenkatesan, T., and Selvaraj, R. (2022). Photocatalytic degradation of methylene blue dye using newly synthesized zirconia nanoparticles. *Environmental Research*, 214(1), 113785. http://doi.org/10.1016/j.envres.2022.113785

Wu, Q., Li, W.-T., Yu, W.-H., Li, Y., and Li Ali, M. (2016). Removal of fluorescent dissolved organic matter in biologically treated textile wastewater by ozonation-biological aerated filter. *Journal of the Taiwan Institute of Chemical Engineers*, 59, 359–364. http://doi.org/10.1016/j.jtice.2015.08.015

Yu, B., He, L., Wang, Y., and Cong, H. (2017). Multifunctional PMMA@Fe$_3$O$_4$@DR magnetic materials for efficient adsorption of dyes. *Mater*, 10, 1–12. http://doi.org/10.3390/ma10111239

Zhang, X., Yang, S., Yu, B., Tan, Q., Zhang, X., and Cong, H. (2018). Advanced modified polyacrylonitrile membrane with enhanced adsorption property for heavy metal ions. *Scientific Reports*, 8, 1260. http://doi.org/10.1038/s41598-018-19597-3.

9 Experimental investigation of orientations effect on 70-watt LED under natural convection

Avadhut R. Jadhav[1,a], Jitendra G. Shinde[1], Gajendra J. Pol[1], Ranjeet S. Mithari[1] and Sujit V. Kumbhar[2]

[1]Bharati Vidyapeeth College of Engineering Kolhapur, Maharashtra, India
[2]Sharad Institute of Technology College of Engineering Yadrav, Ichalkarnji, Maharashtra, India

Abstract

The advancement in the electronic market and utilization of light emitting diode (LED) day by day is rapidly increasing. But the performance of LED is critically affected by high temperatures. Cooling is the main challenge in front of a designer. In the indoor stadium, petrol pumps, like in various commercial applications, LEDs are mounted at different orientations. In this paper, the effect of the orientation of LED on its thermal resistance and cooling performance by natural convection is studied experimentally. LED are generally cooled by the passive cooling technique. i.e., heat sink is present for cooling of LED. Due to ease of manufacturing, rectangular fins are preferred in many heat sinks. As LEDs are mounted at various orientations, it will affect the air-flow circulation, affecting the cooling of the LED heat sink. We observed that at 45° the thermal resistance offered by the same heat sink is less than the other. And at 180° thermal resistance of the heat sink is higher than in other orientations. It is observed that orientation affects the cooling performance of LED. Tilting arrangement is provided to experimental setup carried out experiments carried out at 0, 45, 90, 135, 180, 225, 270, and 315°.

Keywords: Heat sink, light emitting diode, orientation

1. Introduction

In today's world, light emitting diode (LED) is popularly used due to less power consumption. It is a semiconductor device which emits light when the P-N junction applied voltage. Around 7080% of energy turns into heat Jeong et. al. (2015), and 2030% is utilized for light. Moon et al observed AS 7080% of energy is transforms into heat cooling, the key problem is about the cooling of LED. When LED is cooled efficiently the life and efficiency will improve. The rectangular fin heat sink is used in LED for cooling by natural convection. As cooling occurs by passive technique, therefore, material geometrical factors like the length of the fin, the fin width, the fin thickness, the spacing of the fin, and the fin orientation affect the thermal performance of fin.

Many researchers Awasarmol and Pise, Hsieh and Li, work on various heat transfer improving techniques for rectangular fin heat sinks. Sathe, A. and Sanap

[a]arjmesa@gmail.com

DOI: 10.1201/9781003450917-9

and The geometrical parameter plays an important role increase in heat dissipation. Praveen et. al. (2019) took 16 watts LED and analyzed the results of variation of forwarding current on case junction temperature of the LED. They fabricated four LED heat sinks by electric discharge machining. Lin et al.. observed that surface area off in and fin geometrical parameters are the main reason for better heat dissipation. It is noticed that parallel fin heat sink is the optimum heat sink for LED because it has a high surface area which reduces junction temperature and improves the luminous output of the LED.

Haowengong et. al., used an experimental, numerical approach to study several fins, fin height, heat flux, synergy angle, and radiance on thermal resistance and heat transfer. Jeong et. al. (2015) proposed various cooling techniques to improve poor ventilation and heat dissipation of LED heat sinks. They provide an opening in the fin base and fin to improve air circulation, enhancing the LED module's cooling performance. According to them, the total thermal resistance at 180° was less than 0°.

Qie Shen and daming conducted an experimental study on inclination effects on free convection heat transfer rectangular fin heat sink of for eight different angle like 0, 45, 90, 135, 180, 225, and 315°. According to the result it is observed that concentrated fin array are more sensitive to inclination. The exponent of Rayleigh number m is one and same for 45°, 135°, 225°, and 315° orientations. The orientation effect factor η is defined for each orientation, based on which the heat dissipation performances of 135°, 225°, and 315° are 99%, 76%, and 91%, respectively.

LEDs are generally used in domestic and commercial applications like household and streetlamps. Indoor stadiums, petrol pumps, malls. In these applications, LEDs are mounted at various inclinations. In this paper effect of orientation on LED heat sink transfer is studied experimentally at 0, 45, 90, 135, 180, 225, 270, and 315°.

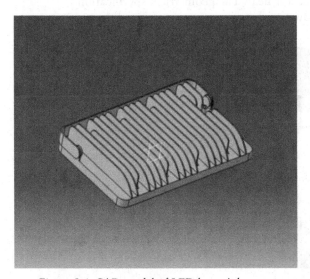

Figure 9.1 CAD model of LED heat sink

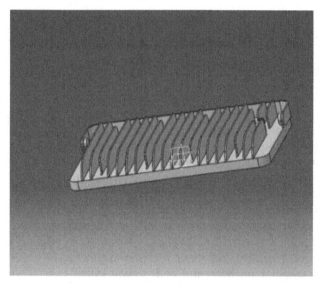

Figure 9.2 CAD model of LED heat sink

Experimental test rig

The experimental setup is designed and developed to take trial on LED heat sink under free convection for various inclinations. The geometrical factor of the heat sink is keeping the same. The main objective of the experiments is to get the best orientation and to utilize the optimum area under mode of natural convection. A 70 watt LED heat is taken for experimentation. The setup is manufactured in such a way that experiment was carried out at various orientations. The thermo-couple is used to record the temperature at multiple locations. For calculation, 80% of energy is assumed to be utilized for heat. The geometrical specification of the LED heat sink is mentioned below.

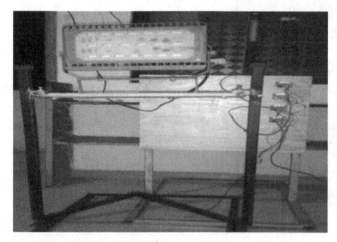

Figure 9.3 Experimental setup

Figure 9.4 LED heat sink

Figure 9.3 shows the experimental setup for recording 70 watts LED heat sink temperature at various inclinations by natural convection. The tilting mechanism is shown in Figure 9.3 to get multiple orientation angles. LED of 70 watts is shown in Figure 9.4 also CAD model of LED is shown in Figures 9.1 and 9.2. The specification of LED and geometrical detail of LED and heat sink is given below:

Fin spacing = 14 mm
Fin length = 167 cm
Fin thickness = 2 mm
Fin height = 27 mm
No. of fin = 20

Assumption:

Fin has a rectangular cross-section.

Temperature is recorded at a steady state.

Fin is assumed to continue, i. e., the discontinuity present in fin for connection purposes, and other purposes are neglected.

Following experimental procedure is followed.

1. The temperature recorded at steady state.
2. The material of fin is homogeneous.
3. Consider the fine array isothermal.
4. Consider heat transfer coefficient is constant and uniform over entire area of heat sink.
5. Maintain uniform local temperature
6. Calculate the experimental result after achieving a steady state
7. Internal heat generation is negligible.

All the temperatures of thermocouples T1 to T3 and Ta are measured after steady-state conditions.

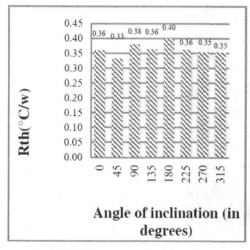

Graph 9.1 Thermal resistance vs angle of inclination

Result and discussion

In this experimental work, experimentation is carried out on 70-watt LED lamp, as shown in fig no. At various locations of the LED heat sink temperature is recorded (at the bottom of the fin, at the top of the fin, at mid of fin). The thermal resistance of LED heat sink calculated for 0, 45, 90, 135, 180, 225, 270, 315. Nine experiments are carried out. The time taken by the heat sink to get to a steady state, from ambient temperature as an initial temperature was on average 25 min. The equivalent thermal resistance between the junction and ambient temperature R_{th} is the criteria that characterizes the effectiveness of a heat sink.

$$Rth = \frac{Tavg - Ta}{Q}$$

Where,
Q = Heat input
v = Voltage
i = Current
T1, T2, T3 = Temperature at the bottom of the fin, at mid of fin and at the top of the fin respectively.
T_{avg} = Average surface temperature of fin
Ta = Ambient air temperature
R_{th} = Thermal resistance between junction and external environment
Table 9.1 shows that at an angle of 45°, the thermal resistance offered by the heat sink is less. So, at 45°, the heat dissipation is more than in other orientations, and at 180°, thermal resistance is maximum compared to other orientations. Graph 9.1 also shows the thermal resistance for various orientations. Also Graph 9.2 shows temperature at the base of the fin, center of fin, and top of fin vs. angle of inclination.

Table 9.1 Table of orientation and thermal resistance

Orientation(ϕ)	0°	45°	90°	135°	180°	225°	270°	315°
Rth (	0.36	0.33	0.38	0.36	0.40	0.36	0.35	0.35

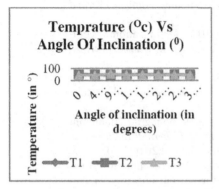

Graph 9.2 Temperature vs angle of inclination

From Graph 9.2, it is clear that temperature decreases as the angle of inclination increases from 0 to 315° except at 90°. At 90°, fin is normal to the direction of fluid flow. Hence, top surface temperature changes drastically at this orientation. For this reason, it is observed that LEDs are normally mounted other than at 90°.

Conclusion

An experimental investigation of 70-watt of light emitting diode (LED) heat sink array was carried out at various directions. From the result it is found that at 45°, the thermal resistance is less and at 180°, the thermal resistance is more. It indicates that airflow is more at 45° more which shows cooling performance of heat sink at 45° is more than other orientations. At 180° the cooling is less as compared to other orientations. So, it can be concluded from the paper that at 45° direction, cooling performance is efficient as compared to other orientations like 0, 90, 135, 180, 225, 270, and 315°.

References

Awasarmol, U. V. and Pise, A. T. (2018). Experimental Study of Heat Transfer Enhancements from Array of Alternate Rectangular Dwarf Fins at Different Inclinations. *Journal of The Institution of Engineers (India): Series C*, 125–131.

Cucumo, M., Ferraro, V., Kaliakatsos, D., and Marinelli, V. (2017). Theoretical and experimental analysis of the performances of a heat sink with vertical orientation in natural convection. *International Journal of Energy and Environmental Engineering*. 8, 247–257.

Dheepan Chakravarthii, M. K., Shanmugan, S., and Mutharasu, D. (2016). Estimation of junction temperature and thermal resistance of LEDs mounted on different heat sinks by numerical simulation and thermal transient measurement. *International Journal of Engineering Trends and Technology (IJETT)*, 42(5), 241–251.

Hsieh, C. C. and Li, Y. H. (2015). The study for saving energy and optimization of LED street light heat sink design. *Hindawi Publishing Corporation Advances in Materials Science and Engineering*, 2015, Article ID 418214, 5.

Jeong, M. W., Jeon, S. W., and Kim, Y. (2015). Optimal thermal design of horizontal fin heat sink with a modified opening model mounted on led module. *Applied Thermal Engineering*, 91, 105–115.

Lin, S.-Q., Shih, T. M., Ru-Gin Chang, R., and Chen, Z. (2016). Optimization of cooling effects with fins of variable cross-sections. *Numerical Heat Transfer, Part A: Applications*, 69(8), 850–858.

Min Woo Jeong, Seung Won Jeon, Yongchan Kim, (2015). Optimal thermal design of horizontal fin heat sink with a modified opening model mounted on Led module, *Applied thermal engineering*, 91, 105–115.

Moon, S.-H., Park, Y.-W., and Yang, H.-M. (2016). A single unit cooling fins aluminum flat heat pipe for 100W socket type COB LED lamp. *Applied Thermal Engineering*, 126, 1164–1169. doi: http://dx.doi.org/10.1016/j.applthermaleng.2016.11.077.

Pankaj, S., Bhosle, S., Kulkarni, K., and Joshi, S. (2018). Experimental investigation of heat transfer by natural convection with perforated pin fin array. *In 2nd International Conference on Materials Manufacturing and Design Engineering. Procedia Manufacturing*, 20, 311–317.

Praveen, A. S., Jithin, R., Naveen Kumar, K., and Mathew, B. (2019). Analysis of Thermal and Optical Characteristics of Light Emitting Diode on Various Heatsinks. *International Journal of Ambient Energy*, 1–15. DOI:10.1080/01430750.2019.156 7585.

Qie Shen, Daming Sun, YaXu, TaoJin, Xu Zhao, (2014). Orientation effects on natural convection heat dissipation of rectangular fin heat sinks mounted on LEDs, *International journal of Heat and mass Transfer*, 75, 462–469.

Sathe, A. and Sanap, S. (2020a). Free convection heat transfer analysis of slitted fin heat sink of vertical orientation using CFD. *International Journal of Ambient Energy*, 2662–2672. DOI: 10.1080/01430750.2020.1758785.

Sathe, A. and Sanap, S. (2020b). Experimental analysis of effect of slitted rectangular fins on heat sink under natural convection heat transfer. *International Journal of Ambient Energy*, 2842–2849. DOI:10.1080/01430750.2020.1778083.

Shen, Q., Sun, D., Xu, Y., Jin, T., and Zhao, X. (2014). Orientation effects on natural convection heat dissipation of rectangular fin heat sinks mounted on LEDs. *International Journal of Heat and Mass Transfer*, 75, 462–469.

Tang, Y., Lin, L., Zhang, S., Zeng, J., Tang, K., Chen, G., and Yuan, W. (2017). Thermal management of high-power LEDs based on Integrated heat sink with vapor chamber. *Energy Conversion and Management*, 151, 1–10.

10 Analysis of a mathematical model on translocation of prey for conservation and biodiversity

Shifa Goyal[1,a] and Nayna Kadam[2]

[1]Department of Mathematics, Maharaja Ranjit Singh College of Professional Sciences, Indore, India

[2]Department of Applied Mathematics, Shri Vaishnav Institute of Technology and Science, Indore, India

Abstract

In South Africa black rhino population was translocated as implementation of national rhino conservation policy. The management of Kanha National Park took conservation measures and successfully translocated 58 Barasingha to Satpura Tiger Reserve. Motivating from this, we study a logistic growth prey-predator model with Holling type-II functional response to study the translocation of prey. Considered model is analyzed and local and global stability conditions for equilibrium points are derived. Numerical examples are taken to illustrate the analytical results.

Keywords: Holling type II functional response, logistic growth, predator, prey, translocation

Introduction

Conservation for biodiversity is important for the present as well as for the generations to come. Sometimes translocation is done to conserve species from extinction (Emslie, 2007) and for biodiversity (Chauhan and Shukla, 2017). The effect of marine reserves in biological conservation of depleted fishing stocks was studied (Ghosh et. al., 2017) and there are many studies on populations living in a confined habitat like (Hearne and Swart, 1991) studied a nonlinear differential equation model on translocation strategies to increase the rhino population in South Africa. In an elementary study based on classical Lotka-Volterra model (Fay and Greeff, 1999) field data was used to study the effect of cropping of the predator population in Kruger National Park in South Africa. In a study (Kar et. al., 2010) both prey and predator species harvested and the effect on the dynamics of the model was analyzed. When predators are harvested then a prey-predator model (Chakraborty et. al., 2012) may show change in stability, oscillations or deterministic extinction. Growth rate of an endangered population can be optimized (Aldila et. al., 2015) through careful implementation of well-planned translocation strategies. The effect of culling on predator population in long duration was also studied (Mickens et. al., 2016). Four different prey predator models (Jha and Ghorai, 2017) with Holling type-II functional response (Holling, 1965) incorporated with

[a]shifamanishgoyal@gmail.com

DOI: 10.1201/9781003450917-10

constant harvesting of prey or predator were analyzed. The effects due to habitat complexity on wildlife population were studied (Pathak, 2018). In a study by Saikia et. al. (2020) the conditions to increase rhino population in Kaziranga National Park under the complexities of poaching, predation and infighting were derived. Inspired from the above studies a mathematical model on the interaction between prey and predator is discussed with the objective to understand how the change in rate of translocation of prey affects prey population over long duration.

The model equations

We modify the model equations given by Aldila et. al. (2015):

$$\frac{dx}{dt} = r_1 x \left(1 - \frac{x}{K_1}\right) - \frac{axy}{1 + bx} - (P + d + R)x|$$

$$\frac{dy}{dt} = r_2 y \left(1 - \frac{y}{K_2}\right) + \beta \left(\frac{axy}{1+bx}\right)$$

$$(1)$$

where, x is the prey population density at any time t; y is the population density of predator at any time t; r_1 is the intrinsic growth rate of prey; r_2 is the intrinsic growth rate of predator; K_1 is the capacity of environment to sustain prey population; K_2 is the capacity of environment to sustain predator population; a is the capture rate on prey; b is the half saturation level; P is the human poaching rate; d is natural death rate of prey; translocation rate of prey is R and β is the efficiency of the predator to convert prey biomass into fertility.

Basic results

It can be easily seen that if variables (x, y) and the parameters are non-negative then right-hand side of both the equations of model (1) are smooth functions, hence existence and uniqueness properties hold in the positive quadrant. Both the prey and predator follow logistic growth rate. We consider the predator a generalist predator because it has other food sources in good quantity. The first equation of model (1) can be written as:

$$\frac{dx}{dt} \leq r_1 x \left(1 - \frac{x}{K_1}\right)$$

which is of logistic form and consequently $x(t)$ is asymptotically less than or equal to K_1 for all t if $x(0) > 0$. From second equation of model (1) it follows that if $y(0) > 0$ then $y(t) > 0$ for all t.

Boundedness of all positive solutions

Theorem 1. All solutions $x(t)$, $y(t)$ of model (1) which start in R_2^+ are uniformly bounded.

Proof. Let $W = x(t) + \frac{1}{\beta}y(t)$ with it's time derivative:

$$\frac{dW}{dt} = \frac{dx}{dt} + \frac{1}{\beta}\frac{dy}{dt}$$

$$= r_1 x\left(1 - \frac{x}{K_1}\right) - \frac{axy}{1+bx}$$
$$- (P + d + R)x$$
$$+ \frac{r_2}{\beta}y\left(1 - \frac{y}{K_2}\right)$$
$$+ \frac{axy}{1+bx}.$$

For each $\eta > 0$ we have

$$\frac{dW}{dt} + \eta W = -r_1\frac{x^2}{K_1} + r_1 x$$
$$- (P + d + R)x$$
$$+ \frac{r_2}{\beta}y\left(1 - \frac{y}{K_2}\right)$$
$$+ \eta x + \frac{\eta y}{\beta}$$

$$\leq Q$$

Where $Q = K_1\frac{[(r_1+\eta)-(P+d+R)]^2}{4r_1} + K_2\frac{[r_2+\eta]^2}{4r_2\beta}$.

On applying differential inequality (Birkhoff and Rota, 1962) we get:

$$0 \leq W(x, y) \leq \frac{Q}{\eta} + \frac{W(x(0), y(0))}{e^{\eta t}}.$$

Letting $t \to \infty$ the above inequality yields $0 \leq W \leq \frac{Q}{\eta}$. Thus, all solutions of the model (1) lie in the region:

$$Q_1 = \left\{(x, y): 0 \leq W \leq \frac{Q}{\eta} + \epsilon; for \; \epsilon \right.$$
$$\left. > 0\right\}.$$

Equilibrium points

The model (1) possess:

1. Trivial equilibrium point $E_0(0, 0)$ which always exists;
2. Boundary equilibrium points:

- $E_1 (\hat{x}, 0)$ exists if $r_1 - (P + d + R) > 0$, Where $\hat{x} = K_1 \frac{r_1 - (P+d+R)}{r_1}$;
- $E_2 (0, \check{y})$, where $\check{y} = K_2$.

3. Interior equilibrium point $E_3(\tilde{x}, \tilde{y})$

Here, $\tilde{x}$ are $\tilde{y}$ the positive solutions of the following equations:

$$r_1 \left(1 - \frac{x}{K_1}\right) - \frac{ay}{1 + bx} - (P + d + R) = 0;$$

$$r_2 \left(1 - \frac{y}{K_2}\right) + \beta \left(\frac{ax}{1 + bx}\right) = 0;$$

Where, $\tilde{y} = \frac{K_2}{r_2} \left[r_2 + \frac{a\beta \tilde{x}}{1 + b\tilde{x}}\right]$ and $\tilde{x}$ satisfies following equation:

$$A\tilde{x}^3 + B\tilde{x}^2 + C\tilde{x} + D = 0 \quad (2)$$

Where,

$$A = \frac{r_1 b^2}{K_1};$$

$$B = \left[\frac{2br_1}{K_1} - b^2 [r_1 - (P + d + R)]\right];$$

$$C = \left[\frac{r_1}{K_1} + K_2 \left(ab + \frac{a^2 \beta}{r_2}\right) - 2b[r_1 - (P + d + R)]\right];$$

$$D = [aK_2 - [r_1 - (P + d + R)]].$$

According to Descartes rule of signs (Jameson, 2006) if any of the following conditions hold good then equation (2) will possess a unique positive solution:

1. $B < 0, C < 0, D < 0$;
2. $B > 0, C < 0, D < 0$;
3. $B > 0, C > 0, D < 0$.

Condition (i) holds if

$$(P + d + R) < r_1 - C_1,$$

Where

$$C_1 = \max \left[\frac{2r_1}{bK_1}, \frac{1}{2b}\left[\frac{r_1}{K_1} + K_2(ab + \frac{a^2 \beta}{r_2})\right], aK_2\right].$$

Condition (ii) holds if

$$r_1 - \frac{2r_1}{bK_1} < (P + d + R) < r_1 - C_2,$$

Where

$$C_2 = max\left[\frac{1}{2b}\left[\frac{r_1}{K_1} + K_2\left(ab + \frac{a^2\beta}{r_2}\right)\right], aK_2\right].$$

Condition (iii) holds if

$$r_1 - C_3 < (P + d + R) < r_1 - aK_2,$$

Where

$$C_3 = max\left[\frac{2r_1}{bK_1}, \frac{1}{2b}\left[\frac{r_1}{K_1} + K_2\left(ab + \frac{a^2\beta}{r_2}\right)\right]\right].$$

Unique positive solution of $\tilde{x}$ obtained from equation (2) will give the interior positive solution of $\tilde{y}$. Thus, we see that under certain parametric restrictions model (1) has different interior equilibria.

Stability analysis

The stability of different equilibrium points of model (1) is discussed with the help of variational matrix and Routh-Hurwitz criterion (May, 2019; Walter and Peterson, 2010).

The variational matrix of model (1) is as below

$$J = \begin{bmatrix} r_1 - (P + d + R) - \frac{2r_1 x}{K_1} - \frac{ay}{(1 + bx)^2} & \frac{-ax}{1 + bx} \\ \frac{a\beta y}{(1 + bx)^2} & r_2 - \frac{2yr_2}{K_2} + \frac{a\beta x}{1 + bx} \end{bmatrix}$$

Case I

The variational matrix of model (1) at $E_0(0, 0)$ is

$$J_{(0,0)} = \begin{bmatrix} r_1 - (P + d + R) & 0 \\ 0 & r_2 \end{bmatrix}$$

$J_{(0,0)}$ is unstable as both the eigen values are positive.

Case II

At $E_1(\hat{x}, 0)$ the variational matrix of model (1) is

$$J_{(\hat{x},0)} = \begin{bmatrix} -\Omega & \dfrac{-a\Omega K_1}{r_1 + b\Omega K_1} \\ 0 & r_2 + \dfrac{a\beta\Omega K_1}{r_1 + b\Omega K_1} \end{bmatrix}$$

Where $\Omega = r_1 - (P + d + R)$ and $\hat{x} = \dfrac{\Omega K_1}{r_1}$. $J_{(\hat{x},0)}$ is stable if

$$r_1 - (P + d + R)$$
$$< -\dfrac{r_1 r_2}{K_1(br_2 + a\beta)}.$$

Case III

The variational matrix of model (1) at $E_2(0, \breve{y})$ is

$$J_{(0,\breve{y})} = \begin{bmatrix} r_1 - (P + d + R) - aK_2 & 0 \\ a\beta K_2 & -r_2 \end{bmatrix}$$

$J_{(0,\breve{y})}$ will be stable if

$$r_1 - (P + d + R) > aK_2$$

Case IV

The variational matrix of model (1) at $E_3(\tilde{x}, \tilde{y})$ is

$$J_{(\tilde{x},\tilde{y})}$$
$$= \begin{bmatrix} \dfrac{-r_1\tilde{x}}{K_1} + \dfrac{ab\tilde{x}\tilde{y}}{(a + b\tilde{x})^2} & \dfrac{-a\tilde{x}}{1 + b\tilde{x}} \\ \dfrac{a\beta\tilde{y}}{(1 + b\tilde{x})^2} & \dfrac{-r_2\tilde{y}}{K_2} \end{bmatrix}$$

The characteristic equation is:
$\lambda^2 + M_1\lambda + M_2 = 0$, where

$$M_1 = \dfrac{r_1\tilde{x}}{K_1} + \dfrac{r_2\tilde{y}}{K_2} - \dfrac{ab\tilde{x}\tilde{y}}{(1 + b\tilde{x})^2}$$

$$M_2 = \dfrac{r_1 r_2\tilde{x}\tilde{y}}{K_1 K_2} - \dfrac{ab\tilde{x}\tilde{y}^2 r_2}{(1 + b\tilde{x})^2 K_2} + \dfrac{a^2\beta\tilde{x}\tilde{y}}{(1 + b\tilde{x})^3}$$

Table 10.1 below is the summary of the stability of interior equilibrium points along with conditions is given:

Table 10.1: Summary

$M_1 > 0$, $M_2 > 0$	'$E_3(\tilde{x}, \tilde{y})$ is either a stable node or stable spiral'
$M_1 < 0$, $M_2 < 0$	'$E_3(\tilde{x}, \tilde{y})$ is either an unstable node or an unstable spiral'
$M_2 < 0$	'$E_3(\tilde{x}, \tilde{y})$ is a saddle point'
$M_1 = 0$	'$E_3(\tilde{x}, \tilde{y})$ is some limit cycle'

The global stability of the interior equilibrium point $E_3(\tilde{x}, \tilde{y})$ in the interior of R_2^+ is investigated by using Lyapunov direct method (Smale, 1974), as shown in the following theorem.

Theorem 2. The interior equilibrium if exists is locally asymptotically stable, provided the following condition holds:

$$(1 + bx) > \frac{abK_1\tilde{y}}{r_1(1 + b\tilde{x})}$$

$$\frac{dV}{dt} = L_1\left(\frac{x - \tilde{x}}{x}\right)\frac{dx}{dt} + L_2\left(\frac{y - \tilde{y}}{y}\right)\frac{dy}{dt}$$

$$= L_1(x - \tilde{x})\left[r_1\left(1 - \frac{x}{K_1}\right) - \frac{ay}{1 + bx} - (P + d + R)\right]$$

$$+ L_2(y - \tilde{y})\left[r_2(1 - \frac{y}{K_2}) + \frac{a\beta x}{1 + bx}\right]$$

$$= -\left[\frac{r_1 L_1}{K_1} - \frac{abL_1\tilde{y}}{(1 + bx)(1 + b\tilde{x})}\right](x - \tilde{x})^2 - \frac{r_2 L_2}{K_2}(y - \tilde{y})^2$$

$$+ (a\beta L_2 - abL_1\tilde{x} - aL_1)\frac{(x - \tilde{x})(y - \tilde{y})}{(1 + bx)(1 + b\tilde{x})}$$

Taking $L_1 = 1$ and $L_2 = \frac{(1 + b\tilde{x})}{\beta}$ we get;

$$\frac{dV}{dt} = -\left[\frac{r_1}{K_1} - \frac{ab\tilde{y}}{(1 + bx)(1 + b\tilde{x})}\right](x - \tilde{x})^2 - \frac{r_2(1 + b\tilde{x})}{\beta K_2}(y - \tilde{y})^2$$

If $(1 + bx) > \frac{abK_1\tilde{y}}{r_1(1 + b\tilde{x})}$, then $\frac{dV}{dt}$ is negative definite and consequently $E_3(\tilde{x}, \tilde{y})$ is locally asymptotically stable.

Numerical examples

For simulation we take some hypothetical values which satisfy analytic conditions and carry out numerical simulation as shown in Figure 10.1 with the help of MATLAB software package.

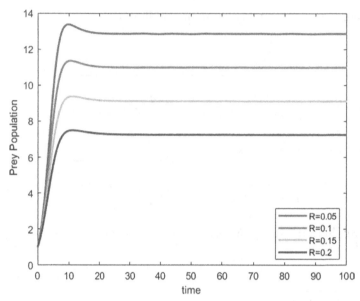

Figure 10.1 Plot of prey population for different values of R

We take parameter values as r_1 = 0.840, K_1 =35, a = 0.086, b = 0.0017, r_2 = 0.499, K_2 = 5, P = 0.01, d = 0.005, β = 0.05, and plot the graph for R = 0.050, R = 0.1, R = 0.15 *and R* = 0.2

Conclusion

We analyzed a deterministic model consisting of two ordinary differential equations. The stability of dynamical system is analyzed and conditions for stable node or spiral and limit cycle are derived mathematically. Numerical simulation is carried out to study the impact of increase in rate of translocation in view of ecology. The prey population is found to be stable at R = 0.050 but when dynamics of model are studied at R = 0.1, R = 0.15, and R = 0.2, then prey population shows continuous decline. It should be kept in mind that growth rates of populations need not remain stable over such a long period of time hence habitat management and a good understanding of ecology of species we can look ahead for a stable state of coexistence at R = 0.050.

Acknowledgment

The first author acknowledges the computational facilities available in the Bhaskaracharya Mathematics Laboratory which is developed through DST-FIST Project (File No.:SR/FST/MSI/2018/26) at the Department of Mathematics, IIT Indore. The author is also grateful to Dr. Bapan Ghosh and Rajni (PhD Scholar), Department of Mathematics, IIT Indore for their help and comments to improve the content.

References

Aldila, D., Hutchinson, A. J., Woolway, M., Owen-Smith, N., and Soewono, E. (2015). A mathematical model of black rhino translocation strategy. *Journal of Mathematical and Fundamental Sciences*, 47(1), 104–115.

Birkhoff, G. and Rota, G. C. (1962). Ordinary Differential Equations. United States of America: John Wiley and Sons.

Chakraborty, S., Pal, S., and Bairagi, N. (2012). Predator-prey interaction with harvesting: mathematical study with biological ramifications. *Applied Mathematical Modelling*, 36(9), 4044–4059.

Chauhan, J. S. and Shukla, R. (2017). Barasingha (Rucervus duvaucelii branderi): Conservation at Kanha, Kanha Tiger Reserve. MP Forest Department India.

Emslie, R. H. (2007). Workshop on biological management of black rhino. *Pachyderm*, 31, 83–84.

Fay, T. H. and Greeff, J. C. (1999). Lions and wildebeest: a predator-prey model. *Mathematics and Computer Education*, 33, 106–119.

Ghosh, B., Pal, D., Kar, T. K., and Valverde, J. C. (2017). Biological conservation through marine protected areas in the presence of alternative stable states. *Mathematical Biosciences*, 286, 49–57.

Hearne, J. W. and Swart, J. (1991). Optimal translocation strategies for the black rhino. *Ecological Modelling*, 59(3-4), 279–292.

Holling, C. S. (1965). The functional response of predators to prey density and its role in mimicry and population regulation. *The Memoirs of the Entomological Society of Canada*, 97S45, 5–60.

Jameson, G. J. (2006). Counting zeros of generalized polynomials : descartes' rule of signs and laguerre's extensions. *The Mathematical Gazette*, 90(518), 223–234.

Jha, P. K. and Ghorai, S. (2017). Stability of prey-predator model with holling type response function and selective harvesting. *Journal of Applied and Computational Mathematics*, 6, 358.

Kar, T. K., Chakraborty, K., and Pahari, U. K. (2010). A prey-predator model with alternative prey: mathematical model and analysis. *Canadian Applied Mathematics Quarterly*, 18(2), 137–168.

May, R. M. (2019). Stability and Complexity in Model Ecosystems. Princeton University Press.

Mickens, R. E., Harlemon, M., and Oyedeji, K. (2016). Consequences of Culling in Deterministic ODE Predator-Prey Models. arXiv preprint. arXiv:1612.09301. https://doi.org/10.48550/arXiv.1612.09301

Pathak, R. (2018). Depletion of forest resources and wildlife population with habitat complexity: a mathematical model. *Open Journal of Ecology*, 8(11), 579.

Saikia, M., Maiti A. P., and Devi, A. (2020). Effect of habitat complexity on rhinoceros and tiger population model with additional food and poaching in kaziranga national park, Assam. *Mathematics and Computers in Simulation*, 177, 169–191.

Smale, S. (1974). Differential Equations, Dynamical Systems, and Linear Algebra. California: Academic Press.

Walter, G. K. and Peterson, A. C. (2010). Autonomous system. In: The Theory of Differential Equations: Classical and Qualitative. New York: Springer.

11 Performance evaluation of flyash and metakaolin with and without Fibrofor fiber reinforced SCC

Arun Kumar H R[a], S Vijaya, M.V. Renukadevi and Somanath M Basutkar

Visvesvaraya Technological University, Belagavi, India

Abstract

Self-compacting concrete (SCC) has been utilized extensively for pouring concrete in congested structural element reinforcement under challenging casting conditions. For these applications, fresh concrete needs to have a high fluidity and strong cohesion. Flyash (FA), a mineral admixture, is a common pozzolonic ingredient. The addition of FA to concrete extends its technical advantages and features while also lowering environmental pollution. Greater. than FA, the concrete strength at 28 days is increased when metakaolin (MK) is used to partially replace cement. Hence, then the MK, the FA increased the strength of the SCC at 90 days. Numerous micro cracks or larger cracks exist in concrete structures, which can start and spread in response to dynamic loads like impacts, earthquakes, and blasts. As a result, the structural parts will be less stable, and if the spreading crack cannot be halted, it would result in severe disasters. Therefore, a new fire called Fibrofor (FF) fiber (High Grade) is employed in this study to avoid such cracks, and very little research has been done on this fiber. The major goal of this study is to compare the qualities of conventional concrete with those of fiber reinforced self-compacting concrete (FRSCC), which has FF added plus FA and MK in place of cement at a specific percentage. It was investigated that how FF fiber varied by 1, 1.5, and 2.0 kg/m^3, FA by 10, 20, 30, 40 and 50%, and MK by 5 and 10%. The Okamura method was used to create a total no. of 35 mix design, including the control mix. On freshly laid concrete, test like slump=flow, L-box, U-box, and V-funnel" were performed. The specimen underwent tests to determine its compression and split tensile strength at ages 7 and 28 days. The outcomes are compared to those of conventional concrete. The results show that adding an additional 1.0 Kg/m^3 FF fiber and replacing Portland cement with 10% MK, 20% FA gave maximum strength.

Keywords: Fibrofor fiber, flyash, mechanical properties, metakaolin, self-compacting concrete

Introduction

Self-compacting concrete (SCC) fills up heavily reinforced portions without vibrating and has a high resistance to segregation and good deformability. It is simple to put and compact under its own weight. Due to the use of admixtures, SCC's 28 day strength was lesser than that of conventional concrete, according to the findings of compressive strength tests on hardened concrete, although SCC eventually demonstrated promise for higher strength beyond 90 days [Hajime Okamura et al.] Due to the characteristics of SCC, casting concrete using them

[a]arunkumarhr1989@gmail.com

DOI: 10.1201/9781003450917-11

requires less effort and is suitable for use in constructions with intricate or tight sections. Conventional concrete and SCC, on the other hand, have generally low tensile and flexural strength as well as weak resistance to the development of cracks and shrinkage [Rahmat Madandoust et al.] The characteristics of hardened concrete are improved when fibres are added to SCC. The addition of fibres to concrete can prevent unexpected failure, enhance fracture energy, reduce crack breadth, reduce shrinkage, and improve flexural, tensile strength, and toughness [Døssland ÅL, Jansson A]. Due to a higher pozzolanic reaction rate, improved particle packing, and accelerated cement hydration, the MK increases compressive strength at young ages (under 28 days) [E.H.Kadri]. For their experimental study, Mousavi [R.Madandoust] utilised MK in 4 distinct ratios. MK was used in replacement of cement in amounts of 5, 10, 15, and 20 %. They found a considerable increase in early age compressive strength after adding MK. At 28days, the compression strength had grown by 27%. Low water absorption was also achieved by using MK. After comparing several MK replacement amounts for their economic viability, fresh concrete qualities, and the properties of hardened concrete, it was found that 10% is the best replacement level. The best outcomes are obtained when 10% MK is used, which increase. the tension- strength" of the -concrete. However, compared: to regular concrete, -the inclusion of superplasticizer also boosts tensile strength [Shahiron Shahidan et al]. The fresh characteristics of SCC can be considerably enhanced by the addition of fly ash. The workability of SCC increases as FA percentage increases. Within the necessary parameters, viscosity, passing, filling capability, and segregation resistance were accepted. The drying shrinkage of SCC's containing FA was significantly smaller than that of normal concrete because FA decreased the rate of hydration [Jamila M. Abdalhmid et al]. A sustainable concrete known as the high volume fly-ash SCC (HVFA-SCC) is proposed to reduce the carbon footprint caused by cement manufacturing and increase the value-added utilisation of admixtures [Jeffery S. Volz, K.Celik] In this mixing technique, the term "HVFA-SCC" refers to cement replacing at least 50% of the fly ash [11]. According to reports, the HVFA-SCC demonstrated much less chloride ion permeability than traditional concrete, which might increase its toughness[W. Wongkeo et al, P. Dinaka et al]. SCC is used in precast concrete for bridges structures, and other applications. In order to increase the concrete's durability, fibres are added to the mix. These fibres act as crack stoppers by halting the spread of cracks. These fibres are randomly arranged and distributed uniformly. The main goals of adding fibres to concrete are to improve its crack stability and crack prevention, as well as to increase the concrete's surface ductility and energy absorption potential [Jiping BAI et al,]. The goal of the study was to examine the suitability of fibre and mineral admixtures based SCC and to improve the mechanical properties of SCC by adding discrete fibre that was randomly oriented and made of polypropylene. The primary goal of the study is to determine how the fresh and mechanical characteristics of SCC are modified by the incorporation of FA and MK by partially substituting cement. Several tests on freshly laid concrete were conducted to evaluate concrete's capacity f or filling and passing ability through and to determine whether it fell into the SCC group. Testing for crushing and spliting-tension-strength at 7" & 28 days for all fibre reinforced SCC mixes was done to determine the impact of including MK, FA, and Fibrofor fibre into SCC The expan- sion in compressive strength at a specific FA portion can be made sense of by the way that the ideal dose of FA is important to totally fill the pores that are available inside the substantial. The expansion of better cementitious particles speeds up the pozzolanic response with CH. Consequently consolidating FA with SF or GGBS can further develop the strength properties of FA admixed concrete [Dheeresh Kumar Nayak et al,] In tests on Self-compacting concrete with MK and GGBS, crushing strength of SCC was observed to be stronger for MK than GGBS at both the prior and later ages. MK outperformed GGBS in terms of starting strength. The latter strength rise brought about by Ca(OH)2 caused the GGBS to react slowly. The structural behaviour of a self-compacting, often vibrated composite column was tested in a number of ways utilising 15% metakaolin as a cement substitute. Reports claim that SCC composite short columns behave better than CC short columns. Due to the presence of the appropriate mineral admixtures, SCC with a large volume of admixtures might minimise CO2 emissions and the heat of hydration. This would promote environmental sustainability [S.S. Vivek]. SCC is used in precast concrete for important projects including houses, bridges, and other structures. In order to increase the concrete's longevity, fibres are occasionally added to the mix. As crack connectors, these fibres stop the fissures from spreading. These fibres are randomly arranged and distributed uniformly. The primary objectives of incorporating fibres to the concrete mixture mixture are to enhance the SCC's "post-cracking" performance. It has been found, that the toughened properties increase by up to 20% when the amount of Metakaolin increases before declining. The results overwhelmingly favour using Metakaolin in place of cement when producing self-compacting concrete [Kummitha Anji Reddy, et al,]. FA's spherical particles are replaced by ettringite needles in severe environments, which break down due to a pozzolanic reaction and produce a denser binder mix than typical concrete. FA concrete has a comparatively low microcrack rate because of the formation of tobermorite, a powerful hydration product under high pressure and temperature, as well as the thick FA particle packing. [Dheeresh Kumar Nayak, et al,].

Material used

53 grade OPC was used, tested in accordance with IS 12269-1987, and confirmed to be satisfactory. The components' respective G and fineness are 3.15 and 2.37%. FA and MK were used to replace some of the cement. The FA utilised comes from coal-fired power plants as a waste product. Its physical form is white powder, and MK is an artificial pozzolana admixture made from thermally activated common clay and kaolinitic clay. Its hue is grey (blackish). The used FA and MK have G (Sp.Gt) of 2.13 and 2.65, respectively. The M-sand-used as the fine=aggregate has a G (Sp.Gt) of 2.67 and was utilised in line with IS: 2386-1963. It falls under the grading zone that it certifies, which must be consistent throughout the entire work.CA with a G (Sp.Gt) of 2.65and a size of less than 12.5 mm was used. Less than 15% made up the Flakiness and Elongation Index. The super plasticizer (SP) GLENIUM B-233 is used to enhance the freshness of SCC.

Coarse aggregate

Angular aggregates with a particle size of 12 mm that were easily accessible locally were used in the current experimental work. The G (Sp.Gt) and fineness modulus are 2.65 and 6.8, respectively

Fibro for fiber

A breakthrough product called FF fiber was created by Brugg Contec, a Swiss company with a successful engineering track record dating back 30 years. They have also created a variety of fibers, the two most well-known of which being FF and Concrix fibers. Which is:

- High performance fiber.
- Bundled, fibrillated, constant spreading and linking.
- Consists of macro-polypropylene fiber.

The properties of Fibrofor fibre is presented in Table-11.1

Super plasticizers

In this investigation, super plasticizers based on poly-carboxylic ether (Glenium B-233) were used. The properties of super plasticizers, GLENIUM B-233 is presented in Table-11.2.

Mix proportions

Material calculation for 20 liters of concrete

Results and discussion

Fresh property studies

Material calculation for different mix ratio is shown in table 11.3. The quality that distinguishes the SCC's capacity to flow into formwork and fill every

Table 11.1: Properties FF fiber

Form	Fibrillated
Bulk-density	0. 91
Length of fiber	19 mm, allowance-+/ 1.5mm
	38-mm = allowance +/.5%
Acids/alkalis-resistance	Inert
TS	400 MPA
Modulus-of-elasticity	4900 MPA
Temperature at which material softens	1500 C
Diameter of foil	80 micro meters

available space under its own weight is known as the filling or flow ability. The slump, T50 slump, and V-funnels test are used to determine filling ability, while the J-Ring and L-Box tests are used to determine passing ability as shown in table 11.4, 11.5 and 11.6. The findings implied that all SCC blends were well within the parameters outlined in the EFNARC standards. The mix ratio M-13 with 50% FA partial cement substitution and 10% MK had the largest slump flow diameter as shown in fig.11.1, values below that were achieved because to increased viscosity that the mix produced during flow. The values obtained in T500 tests as shown in fig 11.2 that were greater than 2 sec belonged to the VS2 category of Viscosity class-2. The V-funnel flow time ranged from 9 to 9.4 sec as shown in fig 11.3. The L-box test was used to determine the effect of obstruction, and the results ranged from

Table 11.2: Properties of super plasticizer

S No.	Description	Value (As per manufacturer)
1.	Form	Liquid
2.	Color	Brown (light)
3.	G (Sp.Gt)	1.2
4.	Resistance to flow (25 deg centigrade)	150 (cps)
5.	Value of pH	6. to 9
6.	Dosage	500 ml to 1500 ml/Kg of cementitious material
7.	Content of chloride	<0.2%

Table 11.3: Details of mix proportion for 1.0, 1.5, 2.0 Kg/m^3

Mix ID	Type of mix	Water (Lit/m^3)	Metakaolin (Kg)	Flyash (Kg)	Cement (Kg)	Super plasticizer (ml)		
						1	1.5	2.0 (Kg/m^3)
Mix -1	Normal mix	186.0	-	-	11.72	-	-	-
Mix -2	SCC	186.0	-	-	11.72	88	88	88
Mix -3	F SCC	186.0	-	-	11.72	90	91	92
Mix -4	10FA5MK	186.0	0.47	0.85	9.96	93	93	94
Mix -5	20FA5MK	186.0	0.47	1.7	8.8	95	96	97
Mix -6	30FA5MK	186.0	0.47	2.56	7.26	96	98	99
Mix -7	40FA5MK	186.0	0.47	3.4	6.45	97	99	100
Mix -8	50FA5MK	186.0	0.47	4.26	5.3	99	101	103
Mix -9	10FA10MK	186.0	0.93	0.85	9.4	94	95	97
Mix -10	20FA10MK	186.0	0.93	1.7	8.2	97	98	99
Mix -11	30FA10MK	186.0	0.93	2.56	7.03	99	100	102
Mix -12	40FA10MK	186.0	0.93	3.4	5.86	100	102	102
Mix -13	50FA10MK	186.0	0.93	4.26	4.69	101	103	104

0.85 to 0.94 as shown in fig 11.5. The aforementioned L-box test readings were more than 0.8 and fell into the PA2 passing ability class based on the viscosity class of the EFNARC recommend ations. When compared to standard SCC mix, it was shown that the less slump flow spread diameter, rise in T500 time, decrease in J-ring, and increase in V-funnel time. It's possible

Table 11.4: Fresh properties of FRSCC with 1 Kg/m³ fibro for fiber

Mix ID	Slump flow (mm)	T_{50} cm Time (sec)	V-Funnel time (sec)	J-Ring slump (mm)	L-box H ratio
Mix -1	-	-	-	-	-
Mix -2	720	2.5	9	640	0.85
Mix -3	700	4.8	10	600	0.9
Mix -4	720	4.8	9.9	610	0.9
Mix -5	720	4.7	9.7	615	0.88
Mix -6	730	4.7	9.5	630	0.88
Mix -7	740	4.7	9.2	640	0.86
Mix -8	740	4.6	8.9	650	0.86
Mix -9	760	4.7	9.7	615	0.9
Mix -10	770	4.7	9.5	620	0.88
Mix -11	770	4.6	9.2	640	0.88
Mix -12	780	4.6	9.0	660	0.86
Mix -13	790	4.5	8.5	670	0.86
EFNARC min	650	2	8		0.8
EFNARC max	800	5	12		1

Table 11.5: Fresh properties of FRSCC with 1.5 Kg/m³ FF fiber

Mix- ID	Slump flow (mm)	T_{50} cm time (sec)	V-funnel time (sec)	J-Ring slump (mm)	L-box H ratios
Mix -1	-	-	-	-	-
Mix -2	720	2.5	9	640	0.85
Mix -3	680	4.9	11	570	0.92
Mix -4	690	4.9	10.6	570	0.92
Mix -5	700	4.9	10.2	590	0.9
Mix -6	720	4.8	9.8	610	0.9
Mix -7	730	4.8	9.7	630	0.9
Mix -8	740	4.7	9.4	640	0.9
Mix -9	700	4.9	10.4	580	0.92
Mix -10	710	4.8	10	600	0.92
Mix -11	720	4.7	9.8	620	0.9
Mix -12	740	4.7	9.5	640	0.9
Mix -13	760	4.6	9.2	640	0.9
EFNARC min	650	2	8		0.8
(EFNARC) max	800	5	12		1

Table 11.6: Fresh properties of FRSCC with 2.0 Kg/m³ FF fiber

Mix ID	Slump flow (mm)	T_{50} cm time (sec)	V - funnel time (sec)	J Ring slump (mm)	L-box. H ratio
Mix -1	-	-	-	-	-
Mix -2	720	2.5	9	640	0.85
Mix -3	660	5	11	550	0.96
Mix -4	700	4.9	10.9	560	0.96
Mix -5	710	4.9	10.5	570	0.96
Mix -6	730	4.8	10	575	0.94
Mix -7	740	4.8	9.8	590	0.94
Mix -8	750	4.7	9.5	610	0.94
Mix -9	720	4.9	10.5	570	0.96
Mix -10	720	4.9	10.1	570	0.96
Mix -11	740	4.8	9.8	580	0.96
Mix -12	750	4.8	9.7	600	0.94
Mix -13	755	4.7	9.4	610	0.94
EFNARC min	650	2	8		0.8
EFNARC max	800	5	12		1

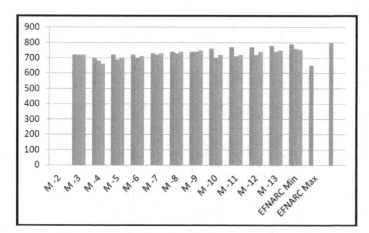

Figure 11.1 Slump flow test on fresh concrete for 1.0, 1.5, 2.0 Kg/m³

that the cause is the mixture's higher yield value and viscosity, as measured by V-funnel test values. Since the produced SCC mixes' viscosity class may be determined by measuring the V-funnel and T500 times in seconds. The findings suggest that the addition of more mineral admixtures boosts the flowing ability. According to tables 11.5, 11.6, and 11.7, the value of slump flow decreased as the FF fibre amount increased, the V-funnel time increased as the amount of fibre increased, the slump flow in the J-ring decreased as the amount of fibre increased, and the L-box H ratio increased as the amount of fibre increased. The slump flow value and J-Ring Slump value are specifically

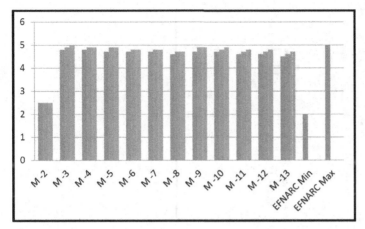

Figure 11.2 T50 (cm) time (sec) test on fresh concrete for 1.0, 1.5, 2.0 Kg/m^3

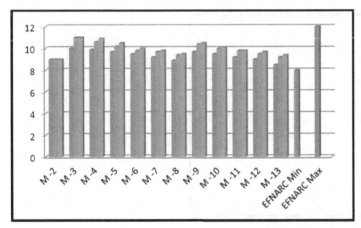

Figure 11.3 V-Funnel time (sec) test on fresh concrete for 1.0, 1.5 and 2.0 Kg/m^3

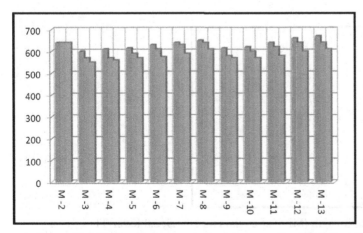

Figure 11.4 J-Ring slump test on fresh concrete for 1.0, 1.5 and 2.0 Kg/m^3

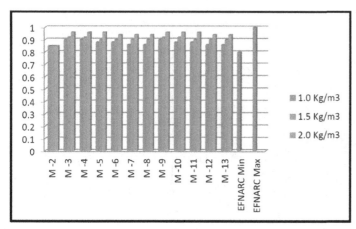

Figure 11.5 L-Box H ratio test on fresh concrete for 1.0, 1.5 and 2.0 Kg/m³

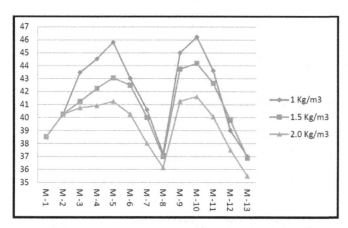

Figure 11.6 Compression strength. (MPA) at 28 days for 1.0, 1.5, 2.0 Kg/m³ FF fiber

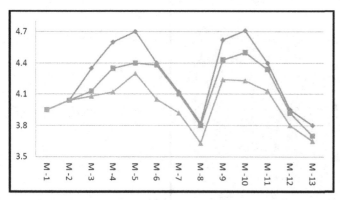

Figure 11.7 Splitting tensile strength (MPA) at 28 days for .1, 1.5, 2.0 Kg/m³ FF fiber

increased by the addition of MK 5% to 10% and FA 10% to 50% to SCC mixes. The cause was MK, which had stronger binding properties due to its creamy nature and might have avoided negative effects like segregation and bleeding during flow.

Mechanical property studies

A cube compressive strength test and cylinder split tensile tests were used to assess the mechanical properties. Tables 11.7, 11.8 & 11.9 display the values for compression and split tensile strength for different fiber dosages. For MK dosages of 5% and 10% as well as FA fluctuations from 10% to 50%, the compression and split tensile strength were calculated. When combined with MK 5% & MK 10%, the ideal FA concentration is 20%, which as increased strength properties shown in table 11.7, 11.8 & 11.9. This study has looked into mix designs with compressive strengths greater than 35 MPa. The mix ratio M-10 for 1.0 kg/m3 Fibrofor fibre has the highest compressive strength among the 36 mix ratios (refer table 11.7, 11.8 & 11.9) measuring 46.22 Mpa. The capacity of fibres to reduce the concentration of stress at the crack tip and control the growth and propagation of cracks may also have contributed to the improvement in compressive strength. Comparing mix ratio M-10 to standard concrete, adding 1.0 kg/m3 of Fibrofor fibre increased compressive strength by 20% as in table 11.7, 14.67% for fiber dose 1.5 kg/m3 & 8.02% for fiber dose 2.0 kg/m3 as shown in table 11.7, 11.8 & 11.9. And also the variation of compression strength for all the mix ratios and for all fiber dose can be seen in fig. 11.6 and 11.8 for 7 & 28 days. The split tensile strength was calculated using a cylinder with dimensions of 300 x 150 mm. The samples were then cast for a further 7 & 28 days. With 20% FA and 10% MK, a considerable increase in strength was seen, reaching 4.71 MPa with the inclusion of fibrofor fibre at

Table 11.7: Mechanical properties of SCC at 28 days with 1Kg/m^3 FF fiber

Mix ID	Type of mix	Compression Strength (N/mm^2)	Splitting tension strength (N/mm^2)
Mix -1	Control mix	38.53	3.95
Mix -2`	SCC	40.26	4.04
Mix -3	FSCC	43.48	4.35
Mix -4	10FA5MK	44.53	4.6
Mix -5	20FA5MK	45.82	4.7
Mix -6	30FA5MK	43.02	4.4
Mix -7	40FA5MK	40.60	4.12
Mix -8	50FA5MK	37.23	3.82
Mix -9	10FA10MK	45.0	4.62
Mix -10	20FA10MK	46.22	4.71
Mix -11	30FA10MK	43.62	4.40
Mix -12	40FA10MK	39.02	3.95
Mix -13	50FA10MK	37.0	3.8

Table 11.8: Mechanical properties of SCC at 28 days with 1.5 Kg/m^3 FF fiber

Mix ID	Type of mix	Compression Strength (MPA)	Splitting tension strength (MPA)
Mix -1	Control mix	38.53	3.95
Mix -2	SCC	40.26	4.04
Mix -3	FSCC	41.24	4.13
Mix -4	10FA5MK	42.25	4.35
Mix -5	20FA5MK	43.05	4.4
Mix -6	30FA5MK	42.5	4.38
Mix -7	40FA5MK	40.02	4.1
Mix -8	50FA5MK	37.0	3.8
Mix -9	10FA10MK	43.75	4.43
Mix -10	20FA10MK	44.18	4.5
Mix -11	30FA10MK	42.65	4.34
Mix -12	40FA10MK	39.8	3.92
Mix -13	50FA10MK	36.89	3.7

Table 11.9: Mechanical properties of SCC at 28 days with 2.0 Kg/m^3 FF fiber

Mix ID	Type of mix	Compression Strength (MPA)	Splitting tension strength (MPA)
Mix -1	Control mix	38.53	3.95
Mix -2	SCC	40.26	4.04
Mix -3	FSCC	40.75	4.08
Mix -4	10FA5MK	40.92	4.12
Mix -5	20FA5MK	41.24	4.3
Mix -6	30FA5MK	40.26	4.05
Mix -7	40FA5MK	38.02	3.92
Mix -8	50FA5MK	36.15	3.63
Mix -9	10FA10MK	41.26	4.24
Mix -10	20FA10MK	41.62	4.23
Mix -11	30FA10MK	40.07	4.13
Mix -12	40FA10MK	37.5	3.8
Mix -13	50FA10MK	35.5	3.65

the recommended dosage of 1.0 kg/m3 i.e.19.24% increase compare to conventional concrete. There is 13.92% increase for 1.5 kg/m3 & 7.1% increase for 2.0 kg/m3 fiber dose as shown in table 11.7, 11.8 & 11.9. A variation of compression & tensile strength can be seen for all fiber dose from fig.11.6 and fig.11.9.

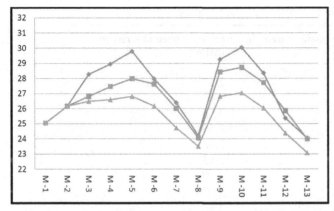

Figure 11.8 Compression/strength (MPA) at 7 days for 1.0, 1.5, 2.0 Kg/m³ FF fiber

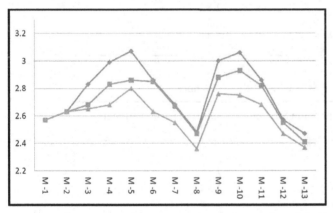

Figure 11.9 Compression strength (MPA) at 7 days for 1.0, 1.5, 2.0 Kg/m³ FF fiber

Table 11.10: Mechanical properties of SCC at 7 days with 1Kg/m³ FF fiber

Mix ID	Type of mix	Compression Strength (N/mm²)	Splitting tension strength (N/mm²)
Mix -1	Control mix	25.04	2.57
Mix -2	SCC	26.17	2.63
Mix -3	FSCC	28.26	2.83
Mix -4	10FA5MK	28.94	2.99
Mix -5	20FA5MK	29.78	3.07
Mix -6	30FA5MK	27.96	2.86
Mix -7	40FA5MK	26.39	2.68
Mix -8	50FA5MK	24.20	2.48
Mix -9	10FA10MK	29.25	3.00
Mix -10	20FA10MK	30.04	3.06
Mix -11	30FA10MK	28.35	2.86
Mix -12	40FA10MK	25.36	2.57
Mix -13	50FA10MK	24.05	2.47

Table 11.11: Mechanical properties of SCC at 7 days_ with 1.5 Kg/m³ FF fiber

Mix ID	Type of mix	Compression Strength (MPA)	Splitting tension strength (MPA)
Mix -1	Control mix	25.04	2.57
Mix -2	SCC	26.17	2.63
Mix -3	FSCC	26.81	2.68
Mix -4	10FA5MK	27.46	2.83
Mix -5	20FA5MK	27.98	2.86
Mix -6	30FA5MK	27.63	2.85
Mix -7	40FA5MK	26.01	2.67
Mix -8	50FA5MK	24.05	2.47
Mix -9	10FA10MK	28.44	2.88
Mix -10	20FA10MK	28.72	2.93
Mix -11	30FA10MK	27.72	2.82
Mix -12	40FA10MK	25.87	2.55
Mix -13	50FA10MK	23.98	2.41

Table 11.12: Mechanical properties of SCC with 2.0Kg/m³ FF fiber

Mix ID	Type of mix	Compression Strength (MPA)	Splitting tension strength (MPA)
Mix -1	Control mix	25.04	2.57
Mix -2	SCC	26.17	2.63
Mix -3	FSCC	26.49	2.65
Mix -4	10FA5MK	26.60	2.68
Mix -5	20FA5MK	26.81	2.80
Mix -6	30FA5MK	26.17	2.63
Mix -7	40FA5MK	24.71	2.55
Mix -8	50FA5MK	23.50	2.36
Mix -9	10FA10MK	26.82	2.76
Mix -10	20FA10MK	27.05	2.75
Mix -11	30FA10MK	26.05	2.68
Mix -12	40FA10MK	24.38	2.47
Mix -13	50FA10MK	23.08	2.37

Conclusion

The investigations indicated above produced the findings that are listed below.

- Slump-flow, T_{500}, V-funnel and L-box characteristics test findings of mineral admixtures-based fiber reinforced self-compacting concrete (FRSCC) mixes have satisfied the limits set by EFNARC recommendations. However, it was implied that the greater cement replacements utilizing metakaolin (MK) and flyash (FA) had decreased the self compacting concrete (SCC) mixes' capacity for filling and passing.

- In both the hardened and fresh states, the *parameters of* a SSC mix containing 10% and 20% MK and 10-50% FA are measured and analyzed. The cube compressive strength increases up to 20%, 14.67%, and 8.02% for Mix -10 ratio at 28 days for fiber content1 kg/m^3 and thereafter decreases for 1.5 and 2.0 kg/m^3 FF fiber, respectively (Tables 11.7-11.9; Figure 11.6). And for Mix-10 ratio toughened qualities were also improved.

- At 28 days, SCC cylinders' maximum splitting tensile strength occurs for a Mix-10 ratio for all three fiber *proportions* 1.0, 1.5, and 2.0 kg/m^3. Fibrofor (FF) fiber, respectively, the minimum splitting tensile strength increases up to 19.24%, 13.92%, and 7.1% for 1.0 Kg/m^3 *fiber content* and thereafter decreases for 1.5 and 2 *kg/m*^{3} (Table 11.7, 11.8 and 11.10; Figure 11.7). The optimal amount of FF fiber addition was found to be 1.0 Kg/m^3.

- The results strongly support the use of MK and FA admixtures in place of cement as well as the addition of 1.0 Kg/m^3. FF fiber leads to increased qualities.

Reference

[1] Abdalhmid, J. M. et al. (2019). Long-term drying shrinkage of self-compacting concrete experimental and analytical investigation. *Construction and Building Materials*, 202, 825–837.

[2] Kemal Celik, Cagla Meral, Mauricio Mancio, P. Kumar Mehta, Paulo J.M. Monteiro. (2014) A comparative study of self-consolidating concretes incorporating high-volume natural pozzolan or high-volume fly ash, *Construction and Building Materials*, 67, 14–19.

[3] Dinakar. P, K.G. Babu, Manu Santhanam, (2008). Durability properties of high volume fly ash self compacting concretes, *Cement and Concrete Composites*. 30(10), November 2008, 880–886.

[4] Døssland ÅL, Aes Lyslo, (2008). Fibre reinforcement in load carrying concrete structures: laboratory and field investigations compared with theory and finite element analysis; ISBN 978-82-471-6924-7. ISBN 978-82-471-6910-0 (printed ver.) ISSN 1503-8181. Thesis for the degree of philosophiae doctor, Norwegian University of Science and Technology.

[5] Jansson A. (2011). Effects of steel fibres on cracking in reinforced concrete. *Chalmers University of Technology*. ISBN 978-91-7385-552-5, ISSN no. 0346-718X, Published by ProQuest LLC (2020) Copyright of the Dissertation is held by the Author, ProQuest Number: 27765630.

[6] Jiping Bai, Stan Wild, Albinas Gailius, (2004). Accelerating early strength development of concrete using metakaolin as an admixture, *Material Science*. 10(4), 338–344.

[7] Kadri E.H Said Kenai, Karim Ezziane, Rafat Siddique, Geert De Schutter, (2011). Influence of metakaolin and silica fume on the heat of hydration & compressive strength development of mortar, *Applied Clay Science*. 53, 704–708.

[8] Madandoust. R S. Yasin Mousavi, (2012). Fresh and hardened properties of SCC containing metakaolin, *Construction and Building Materials*, 35, October 2012, 752-760.

[9] Rahmat Madandoust, Malek Mohammad Ranjbar, Reza Ghavidel, S. Fatemeh Shahabi, (2015). Assessment of factors influencing mechanical properties of steel fiber reinforced self-compacting concrete. *Material & Design,* 83, 284–294.

[10] Nayak D.K, P.P. Abhilash, Rahul Singh, Rajesh Kumar, Veerendra Kumar, (2022). Fly ash for sustainable construction: A review of fly ash concrete and its beneficial use case studies. *Cleaner Materials.* 6, December 2022, 100143.

[11] Okamura H and Masahiro Ouchi, (2003). Self Compacting Concrete, *Journal of Advanced Concrete Technology,* 1(1), 5–15, April 2003/ Copyright © 2003 Japan Concrete Institute.

[12] Reddy. K.A, Gude Rama Krishna, Praveen Kumar Balguri, (2022). Experimental Investigation on the Properties of SCC containing Metakaolin and Polypropylene Fibre. *Materials Today: Proceedings,* 62(Part 6), 3006–3010.

[13] Shahiron Shahidan, Bassam A Tayeh, A.A. Jamaludin, N.A.A.S Bahari, S.S Mohd, N.Zuki Ali and F.S.Khalid, (2017). Physical and mechanical properties of self-compacting concrete containing superplasticizer and metakaolin, *IOP Conf. Series: Materials Science and Engineering,* 271, 012004. doi:10.1088/1757-899X/271/1/012004.

[14] S.S. Vivek. (2022). Performance of ternary blend SCC with ground granulated blast furnace slag and metakaolin, *Materials Today: Proceedings,* 49, 1337–1344. https://doi.org/10.1016/j.matpr.2021.06.422

[15] Jeffery S. Volz, (2012). High-volume fly ash concrete for sustainable construction. *Advanced Materials Research*, 512-515, 2976-2981.

[16] Watcharapong Wongkeo, Pailyn Thongsanitgarn, Athipong Ngamjarurojana, Arnon Chaipanich, (2014). Compressive strength and chloride resistance of self-compacting concrete containing high level fly ash and silica fume, *Materials and Design.* 64, 261–269.

12 Analysis of population growth and its impact on sustainable urban agglomeration

Siddharth[a], Devendra Somwanshi, Vasundhara and Shruti Sharma

Poornima College of Engineering, Jaipur, Rajasthan, India

Abstract

This study aims to analyze the rapid population growth in the growing urban agglomeration of Rajasthan and its impact on the social, economic, and environmental aspects of the region. The study analyses the population growth rates and its impact on the urbanization process in Rajasthan since 2000. The study further investigates the trends in land use, economic activities, and the effect of population growth on the environment. The study also analyses the government policies and programs implemented to address the population growth and its associated problems. Finally, the study proposes sustainable development strategies for the region to cope with population growth and its associated problems. Based on the analysis of data from various sources, the study finds that the population growth rate in Rajasthan has been increasing steadily since 2000, leading to rapid urbanization in the state. The study also finds that the rapid population growth has led to increased pressure on land resources, economic activities, and the environment. The government policies have not been successful in addressing the population growth problem in the region. The study proposes a series of sustainable development strategies for the region to cope with the population growth. These strategies include promoting economic activities, providing access to basic services, developing infrastructure, and ensuring environmental sustainability.

This paper examines the population growth and its impact on the increasing urban agglomeration of Rajasthan, India. With a population of over 68 million people, Rajasthan is the largest state in India and is known for its desert and semi-arid regions. The population growth rate in Rajasthan is higher than the country's average, making it one of the fastest-growing states in terms of population. This paper examines the population growth and its impact on the increasing urban agglomeration of Rajasthan. The study also looks at the implications of population growth for urban development, infrastructure, and services. The paper will analyze the impact of population growth on the urban agglomeration of Rajasthan, by looking at different aspects such as urbanization, migration, and the environment.

Additionally, the paper will discuss the strategies that have been implemented to address the challenges posed by population growth in Rajasthan. Finally, the paper will provide policy recommendations to ensure sustainable development in the face of growing urban agglomeration of Rajasthan.

Keywords: Environment, growth, impact, population, urban

[a]siddharth.choudhary@poornima.org

DOI: 10.1201/9781003450917-12

Introduction

India is one of the most populated countries in the world, and Rajasthan is one of the most populous states in India. The population of Rajasthan has grown exponentially over the past decade, with nearly 68 million people as of 2020. This rapid growth has had a significant impact on the urban agglomeration of Rajasthan and its surrounding areas, as the state has become increasingly urbanized. This paper aims to analyze the population growth of Rajasthan and its impact on the urban agglomeration of the state. The research will include an examination of the factors driving population growth, the impact of population growth on urban development, and the challenges posed by rapid urbanization. The paper will also analyze the opportunities and strategies for sustainable urban development in the context of the growing population of Rajasthan. Finally, the research will suggest possible strategies and solutions to promote sustainable urban growth in the face of population growth.

Rajasthan is one of the most populous states in India and has been witnessing a rapid growth in population, leading to an urban agglomeration of various towns and cities. This rapid growth in population has had a significant impact on the state's economy, infrastructure, and environment. This paper aims to analyze the trend in population growth in Rajasthan and its effect on the growing urban agglomeration. The analysis will consider the factors that have caused this rapid growth in population and the challenges that have been created due to it. It will also discuss the various strategies and policies that can be used to control this population growth and its associated issues. Finally, it will evaluate the efficacy of such strategies and policies to ensure a sustainable growth in population in the state.

Literature review

Urban growth of Rajasthan has been increasing (Kumar, 2017) since the early 2000s due to a number of reasons, including increasing population, economic growth, and infrastructure development.

This growth has had a significant impact on the state's urban agglomeration (Dhankhar and Agarwal, 2011), which is defined as a densely populated area of urban and suburban development (Jain and Dixit, 2011). This paper reviews the literature on the population growth and its impact on the urban agglomeration of Rajasthan.

The literature reveals that Rajasthan's population growth rate has been increasing steadily over the past few decades due to a number of factors, including urbanization, migration, and natural population growth (Singh and Sharma, 2017). According to the 2011 census, the state's population was approximately 68 million, with an annual growth rate of 2.5% (Tiwari, 2013). This population growth has increased the demand for urban housing, leading to a rapid expansion of urban agglomerations. The literature also reveals that this population growth has had a number of impacts on the state's urban agglomeration. Firstly, it has resulted in a higher demand for housing, leading to an increase in the number of slums and informal settlements (Singh and Sharma, 2012). Secondly, the population growth has resulted in a shortage of basic services, such as water, sanitation, and electricity, as well as increased (Tiwari, 2014).

Population growth has a direct impact on the urban agglomeration of any state. Rajasthan is the largest state in India in terms of population and is considered to be one of the most populous states in the country (Kumar, 2015). This article reviews the literature on the analysis of population growth and its impact on the growing urban agglomeration of Rajasthan (Deshpande, 2018). It focuses on the factors influencing population growth and its implications for urbanization in Rajasthan. The literature review on population growth and its impact on urbanization in Rajasthan has identified various factors influencing population growth in the state. The most important factor contributing to population growth in Rajasthan is the high fertility rate (Singh, 2017), which is higher than the national average. This is attributed to the lack of access to modern contraceptives, early marriage and high illiteracy rate among the rural population (Sharma and Rai, 2015).

Other factors influencing population growth in Rajasthan include migration from other states, high birth rate, lack of education and lack of employment opportunities in rural areas. The literature review has also identified the implications of population growth for urban agglomeration in Rajasthan.

Methodology

A systematic approach to analyzing the population growth and its impact on the growing urban agglomeration of Rajasthan would involve the following steps:

1. Data collection: The first step in analyzing population growth and its impact on the growing urban agglomeration of Rajasthan would be to collect data related to the population growth in the region over the years. This data can be collected from secondary sources such as census records, government reports, etc.

2. Data analysis: Once the data has been collected, it needs to be analyzed to identify the trends in population growth and its effects on the urban agglomeration of Rajasthan. This can be done through various statistical techniques such as correlation analysis, regression analysis, etc.

3. Assessment of impact: After analyzing the data, it is important to assess the impact of population growth on the urban agglomeration of Rajasthan. This can be done by looking at various factors such as the availability of resources, infrastructure development, economic development, etc.

4. Recommendations: Based on the assessment, recommendations can be made to address the impact of population growth on the urban agglomeration of Rajasthan (Figure 12.1). These may include policy changes, investments in infrastructure and development, etc.

5. Model development: After the data has been analyzed, the next step is to develop a model that can be used to predict the future population growth and its effect on the urban agglomeration of Rajasthan. This model should take into account the current population trends and the various factors that have influenced the growth.

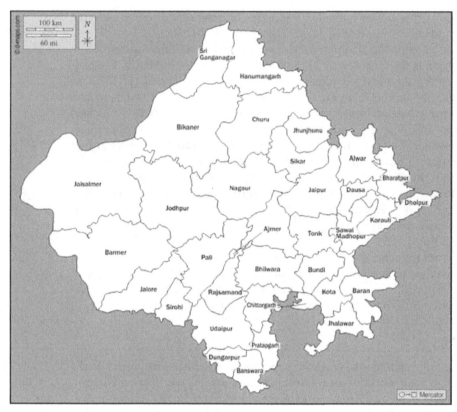

Figure 12.1 Map of Rajasthan

6. Simulation: Once the model is developed, the next step is to simulate the population growth and its effects on the urban agglomeration of Rajasthan. This should include an analysis of the potential for growth in the future, and the potential for negative effects on the environment infrastructure.

Conclusion

In conclusion, it is evident that the rapid population growth in Rajasthan has led to the growth of urban agglomerations, which have in turn impacted the lives of the citizens in the region.

The population growth has led to an increase in demand for resources, services, and infrastructure, and it has also created numerous challenges, such as overcrowding, pollution, and other social and economic issues. Despite these challenges, the growth of Rajasthan's urban agglomerations has also brought with it numerous opportunities, such as increased economic activity, job opportunities, and improved quality of life. The government of Rajasthan must take proactive steps to ensure that the growth of its urban agglomerations is both sustainable and managed properly. This includes promoting good governance and transparency, providing adequate resources and services to meet the needs of its citizens, and taking action to reduce the environmental impacts of its urbanization.

The population growth of Rajasthan is increasing at an alarming rate. This has serious implications for the urban agglomeration of Rajasthan, as it puts a strain on resources and infrastructure, leading to overcrowding, poverty, and uneven development. To sustain the growth and development of Rajasthan, it is essential to focus on improvements in healthcare, education, and employment opportunities, and to ensure that existing resources are efficiently utilized to meet the needs of the population. In addition, the government must take steps to reduce the rate of population growth and promote urban planning that takes into account the needs of current and future generations.

It can be concluded that population growth and its impact on the growing urban agglomeration of Rajasthan have resulted in a number of challenges in terms of infrastructure development, resource utilization, and quality of life for the citizens. The state needs to take a proactive approach to address these issues in order to ensure a sustainable development of the state. It is important to promote policies that foster investment, employment and education opportunities to reduce poverty and to improve the quality of life of its citizens. The effective implementation of such policies will go a long way in providing a better future for the people of Rajasthan.

References

Deshpande, S. (2018). Exploring the urban growth and population change in the state of Rajasthan: a case study of Jaipur. *Urban Studies*, 55(2), 591–607.

Dhankhar, S. and Agarwal, D. (2011). Population growth and its impact on urbanization in Rajasthan. *International Journal of Applied Economics*, 8(1), 63–71.

Jain, S. and Dixit, S. (2011). Population growth and its impact on urban agglomeration in Rajasthan. *Global Journal of Social Sciences*, 10(1), 1–5.

Kumar, P. (2017). Study of population growth and its impact on growing urban agglomeration of Rajasthan. *International Journal of Advance Research and Innovative Ideas in Education*, 3(2), 1215–1219.

Sharma, A. and Rai, A. (2015). Population growth and urbanization in Rajasthan: a geospatial perspective. *Journal of Indian Geographers*, 8(2), 91–106.

Singh, A. (2017). Urban population growth and its impacts on urban economy in Rajasthan, India. *International Journal of Environmental Sciences and Technology*, 14(2), 827–834.

Singh, R. and Sharma, R. (2012). Population growth and its implications on urbanization in Rajasthan. *International Journal of Management and Social Sciences Research*, 1(2), 26–33.

Singh, S. K. and Sharma, S. K. (2017). Population growth and urbanization: a case study of Rajasthan. *International Journal of Social Science and Interdisciplinary Research*, 6(2), 97–105.

Tiwari, R. (2013). Population growth and its impact on urbanization in Rajasthan. *International Journal of Management and Social Sciences Research*, 2(4), 6–15.

Tiwari, J. (2014). Population growth and its impact on urbanization in Rajasthan. *International Journal of Sustainable Development*, 7(3), 282–291.

13 Structural, electric, magnetic and multiferroic properties of $0.4Ba_{0.85}Ca_{0.15}Zr_{0.10}Ti_{0.90}O_3$-$0.6NiFe_2O_4$ multiferroic particulate composite

Sarita Sharma[1,a], Shilpa Thakur[1], V. S. Vats[2], Bharat Bhushan Barogi[2], Govind Singh[2] and N. S. Negi[1]

[1]Himachal Pradesh University, Summer Hill Shimla 171005
[2]Govt. College Dhramshala, Kangra, 176215

Abstract

In this article lead-free multiferroic composites $0.4(Ba_{0.85}Ca_{0.15}Zr_{0.1}Ti_{0.9})O_3 - 0.6NiFe_2O_3$ (0.4BCZT-0.6NFO) were prepared using chemical solution method. Composite is calcinated and sintered at sintering temperature of 1300°C. The structural, microstructure, electrical, magneto-dielectric and magnetic properties of composite BCZT-NFO are primary focus of this research work. Under structural study mixed structural evolution as perovskite phased structure of BCZT and cubic spinal phase structure of NFO without any traceable secondary phase according to the X-ray diffraction (XRD) study. Scanning electron microscopy (SEM) images reveals distribution of small grain of ferroelectric phase (BCZT) within the array of larger grain of ferrites NFO. For lead-free multiferroic composites, magneto-dielectric behavior at room temperature was observed. Long-range ferromagnetic ordering in the sample is revealed by the observed typical ferromagnetic M-H hysteresis loop in the magnetic study. For composites, remnant magnetization (M_r) is 0.338 emu/g and saturation magnetization (M_s) = 7.98 emu/g have been measured. For prepared samples, room temperature and temperature dependent leakage current density behavior also been investigated.

Keywords: Multiferroic, X-ray diffraction, M-H

Introduction

The multiferroic materials are those having both ferromagnetic and ferroelectric properties display at the same time likewise show magnetoelectric (ME) coupling impact. Because of their potential multifunctional applications in novel electronic devices like the miniaturization of memory cells, transducers, sensors, and actuators, magnetoelectric coupling has a significant technological impact (Zhi et al., 2002; Qian et al., 2014; Jian, 2018). As a result, in recent years, research interest in multiferroic composites, particularly designing with lead-free ferroelectrics, has greatly increased. Overcoming the limitations of ferroelectric and ferromagnetic materials in memory device applications is one important aspect of multiferroic system synthesis. As ferroelectric materials have a low coercive field, data writing can be done quickly and easily in ferroelectric-based

[a]sss.sharmasarita@gmail.com

DOI: 10.1201/9781003450917-13

memory devices. However, memory read operation is a concern due to the ferroelectric systems' fatigue characteristics. In a similarly, the absence of a fatigue mechanism in magnetic materials makes it simple to carry out the memory read operation in magnetic memory devices. However, memory write operations are restricted by their high coercive field. The term "direct ME coupling" refers to materials in which the magnetization or polarization are altered with application of external electric or magnetic field directly, whereas "intrinsic magnetization" occurs when electric field is applied. Materials that are multiferroic can also exhibit all three ferroic orders (magnetic, electric, and mechanical) simultaneously. Single-phase ferroelectric materials were the first to exhibit the ME effect, but their use in device applications is uncommon (Nan et al., 2008; Ma et al., 2011). Two factors primarily account for this: the first is that these materials exhibit a ME response well below room temperature and the second is that they produce very little ME effect. However, multiferroic composite materials have a ME coefficient that is much higher than single-phase multiferroics, allowing for a greater degree of freedom ascribed to the availability of numerous materials. Since the applications of magnetic field directly not alter the six polarization in multiferroic composites, but it does modify it indirectly through strain arbitrated ME coupling (Yamasaki et al., 2006). A few examples of lead-free multiferroic composites that have been reported using various synthesis techniques at various temperatures are $NiZnFe_2O_4$-$BaTiO_3$, $NiFe_2O_4$-$BaTiO_3$, $NiFe_2O_4$–PZT and $CoFe_2O_4$–$BaTiO_3$ (Dzunuzovic et al., 2018; Sreenivasulu et al., 2009; Zhai et al., 2004; Mohan and Joy, 2019) in present article we prepared lead-free multiferroic composite $0.4(Ba_{0.85}Ca_{0.15}Zr_{0.1}Ti_{0.9})O_3 - 0.6NiFe_2O_3$ (0.4BCZT-0.6 NFO) using chemical solution method and sintered at 1300°C.

Experimental procedure

$(Ba_{0.85}Ca_{0.15}Zr_{0.1}Ti_{0.9})O_3$ (BCZT) ferroelectric samples were prepared by sol-gel technique. Here the 2-ethylhexanoic acid used as dissolving agent. Barium acetate, calcium acetate, zirconium acetylo acetonaote, and terta-n-butyl orthotitanate are taken. Firstly, appropriate amount barium acetate was dissolved in dissolving agent taken in beaker on heating system at temperature of 80100°C. After dissolving barium acetate calcium acetate is added to precursor followed by suitable quantities of zirconium acetyl acetonoate. After that, precursor heated for an hour to remove any remaining water content. Now to this precursor separately dissolved terta-n-butyl orthotitanate was added to this precursor in a stoichiometric ratio. Finally, the precursor was heated at about 80°C to dry the solution. Then the powder was sintered at 1300⁰C for the time period of five hour. To prepared $NiFe_2O_3$ (NFO) powder, starting chemical nikal(II) nitrate $(Ni(NO_3)_2.6H_2O)$ and iron(III)nitrate $Fe(NO_3)_3.9H_2O$ were used to synthesize nikle-2-ethylhexanoate $(C_7H_{15}COO)_2Ni$ and Iron-3-ethylhexanoate $(C_7H_{15}COO)_3Fe$, respectively. Both these solutions of nikle-2 ethlhexanote and iron-3- ethylenhexonate were mixed in required proportion. Further precursor/solution of NFO was an allowed to dry at 80°C for 5 hr. The dry power was then annealed for five hours at 1300°C. The 4:6 weight proportions of BCTZ and NFO powder were taken for composite. After being thoroughly mixed and ground, it was subjected to structural study. For the electrical measurement pallets of approximate 1 mm was prepared. Further these pallets are subjected to electrical measurements. After that, they were post-sintered at 1300°C for five hours, and silver paste was used for the electric contact on both sides of the pallets.

X-ray diffraction

Figure 13.1 depicts the measured BCZT-NFO composite diffraction patterns from 20 to 70°. The single phase pervoskite-type structure has been used to indexing the ferroelectric phase, ferrite phase XRD patterns are indexed by cubic inverse spinel structure. Whereas, ferroelectric (BCZT) with tetragonal structure has been confirmed by the XRD-derived lattice parameters. Form the XRD reflected line neither the impurity phase nor the additional peak phase peaks are visible in the composite sample, indicating that our composite samples were produced uniformly using the chemical solution method. XRD patterns of composite samples show both ferroelectric and ferrite phase peaks, that rule out the any chemical reaction among the constituent phases during the sintering or while grinding. The Debye-Scherrer formula can be used to extract the sample crystallite size values for the most prominent peaks, which are (110) in the ferroelectric phase and (311) in the ferrite phase. The definition of the Debye-Scherrer formula is:

$$D = \frac{K\lambda}{\beta Cos\theta} \tag{1}$$

Where K, D and β are the shape factor (0.9), crystallite size and full width at half maxima (in radians) respectively.

The wavelength of the Cu-K radiation is 1.54. Table 13.1 displays the samples' observed crystallite size values. The values of lattice parameters of both ferroelectric and ferrite phases are display in Table 13.1. Further the tetragonality ratio (c/a) of ferroelectric phase show the tetragonal behavior of the BCZT (ferroelectric phase) whereas cubic spinal structure is confirmed of the ferrite phase. Structural analysis reveals that the composite samples of BCTZ and NFO ceramics both possess polycrystalline behavior and undergo structural modification that are specific to each phase.

Table 13.1: Structural parameters as crystallite size, lattice parameters and grain size.

Samples	Crystallite size (nm)		Lattice parameter (Å)				Grain size (μm)	
	BCTZ	NFO	NFO	BCTZ			BCZT	NFO
			a	a	c	c/a		
0.8BC TZ-.6NFO	36.45	40.59	8.90	4.00	4.08	1.020	0.46	1.09

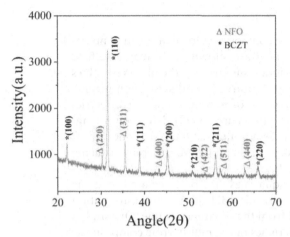

Figure 13.1 XRD reflection lines of composite 0.4BCZT-0.6NFO

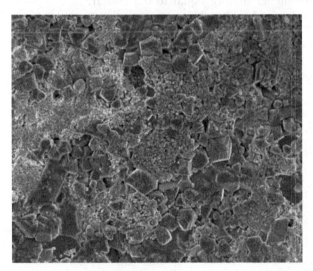

Figure 13.2 SEM images of composite 0.4BCZT-0.6NFO

Scanning electron microscopy

Figure 13.2 depicts composite sample scanning electron micrographs. Due to their different molecular weights, ferrite (NFO) grains appear darker in the SEM images of composite samples, while BCZT grains appear brighter. A similar kind of behavior was also observed by Ramesh et al. (2017). In addition, a well-faced, larger, polygon-shaped grain for NFO and smaller, irregular-shaped grains for BCZT can be seen in the individual micrograph. The presence of a dense microstructure with larger grains in pure CFO phase samples can be attributed to the application of a sufficient sintering temperature, 1300°C, in order to crystallize NFO ceramics. While on another side the ferroelectric phase requires a high sintering temperature for dense microstructure formation and crystallization.

Magnetization properties

Magnetization (M-H) loops of multiferroic composite studied at room temperature at applied magnetic field of 15kOe field are shown to be magnetic field dependent in Figure 13.3. Table 13.2 provides a summary of the observed values for the coercive field (H_c), remnant magnetization (M_r) and saturation magnetization (M_s). It has been noticed that presence of non-magnetic (ferroelectric) phase (BCTZ) in vicinity of the ferromagnetic phase (NFO) deteriorates magnetic properties of the composite samples. As non-magnetic phase hinders magnetic-magnetic interaction among the magnetic (ferrite) grains, thereby reducing the saturation magnetization (Paul et al., 2016).

All ferromagnetic samples have their squareness factor calculated, and the values that were observed are displayed in Table 13.2. In general, information about magnetic domains can be obtained from the squareness value. Our study's results indicate the multi-domain characteristics of the multiferroic composite as squareness values that were observed were less than 0.5. A single domain crystal is characterized by a squareness factor higher than or equal to 0.5 (Saffari et al., 2015). The following equation can be used to determine magnetic moment strength as to Bohr magneton (B).

$$\mu_\beta = \frac{MM_s}{5585} \qquad\qquad (2)$$

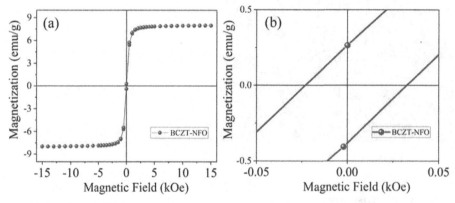

Figure 13.3 M-H hysteresis loop for composite 0.4BCZT-0.6NFO (b) magnified graph of M-H loop

Table 13.2: Parameters as saturation magnetization, remnant magnetization coercive field Bohr magneton as energy of anisotropy.

Samples	M_s(emu/g)	M_r(emu/g)	M_r/M_s	H_c(Oe)	μ_B	K(erg/g)
.4BCTZ - 6NFO	7.98	0.34	0.0 42	28.1 1	0. 31	233.6 4

where 5585, M and M_s are magnetic factor, molecular weight and saturation magnetization respectively. An aspect of magnetostriction in ferromagnetism is calculated as the anisotropy energy. This indicates that the constraint to the ferromagnetic domains to remain unchanged when the magnetic field changes its directions. The equation can be used to determine the value of the anisotropy constant:

$$Anisotropy\ constant(K) = \frac{M_S H_C}{0.96} \qquad (3)$$

0.96, H_c and M_s are the constant factor, coercivity field and is the saturation magnetization respectively. Because anisotropy of materials is influenced by the coercive field, hard ferromagnets possess higher anisotropy than the soft ferromagnetic materials.

Magneto-dielectric properties

Equation can be used to evaluate on of the important behavior of multiferroic composites, which is the magnetocapacitance (MC)/magnetodielectric (MD) behaviors, or the effect of the external magnetic field over the dielectric constant and loss.

$$MD(\%) = \frac{\varepsilon(H) - \varepsilon(H = 0)}{\varepsilon(H = 0)} \qquad (4)$$

Where "(H = 0)" denotes the measurements when there is no external magnetic field and "(H)" denotes measurements of dielectric constant when magnetic field is applied H.

Figure 13.4, which depicts the frequency-dependent variation of dielectric constant at various external magnetic fields, demonstrates magneto-dielectric effect

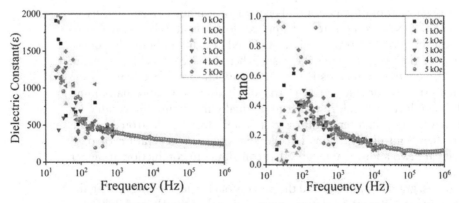

Figure 13.4 Change in dielectric constant and dielectric losses with frequency for composite 0.4BCTZ-0.6NFO

Table 13.3: Parameters such as dielectric constant dielectric loss, magneto-dielectric factors and leakage current density.

Composition	Dielectric constant	Dielectric loss	MD (%)	ML (%)	J (A/cm^2)
.8BCTZ-.6NFO	274.51	0.09	0.086	2.72	1.42 × 10^{-6}

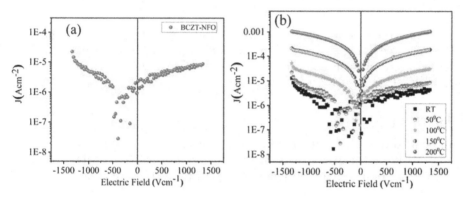

Figure 13.5 Leakage current density of composite 0.6BCTZ-0.4NFO (a) at room temperature and (b) temperature dependent leakage current densities

as an inherent property. It's seen trademark bends that the MD impact is more prominent at a lower frequency while more modest or practically no adjustment of the trademark bend should be visible at a higher frequency. Table 13.3 displays percentile values of the calculated magnetodielectric effect for composite samples at frequency of 1 MHz.

Current density behavior

Figure 13.5(a) show room temperature, behavior of leakage current density of multiferroic BCTZ-NFO composite with applied electric field. When sample subjected to both forward and reverse bias, BCZT-NFO exhibits butterfly curves. The mobility of the charge carrier increases when an electric field is applied. Figure 5(b) depicts the characteristic curves of temperature dependent leakage behavior. With rising temperature free charge of all kinds of valance band jump or excited to conduction band, resulting in gradually increasing the conductivity with temperature. High-temperature responsible for the electron hopping between trivalent and divalent ions like Fe^{3+}/Fe^{2+} and the generation of oxygen vacancies are two additional factors that contribute to conduction at higher temperatures.

Further grains, grain boundaries and the process of charge hopping among the same cations owing different oxidation states (Fe^{2+}/Fe^{3+}, O^{2+}/O^{3+}, and Ti^{3+}/Ti^{4+}), charge mobility and charge percolation all have an impact on the conduction mechanism in multiferroic composites (Bai et al., 2007; Rather et al., 2018). Values of current density measured at 500 V/cm are tabulated in Table 13.3.

Conclusion

Here we effectively synthesis lead free ferroelectric phase (BCZT) by using sol-gel technique and NFO ferrite using metallo-organic method course. Inverse cubic spinal structure for the ferrite (NFO) phase and perovskite structure symmetry for BCZT are confirmed by XRD analysis. Reflection peaks in the composite samples corresponded to both phases without the formation of a secondary phase, indicating that there was no interaction between their phases while synthesizing or sintering. For composite samples, the microstructural investigation reveals a variety of morphologies. For composite samples, magneto-dielectric study reveals a significant variation in the value of the dielectric constant with the applied dc magnetic field. The presence of long-range ferromagnetic ordering in composite is shown by peculiar ferromagnetic M-H hysteresis loops. The leakage current density of sample strongly depends on the temperature. Conduction at lower temperatures is caused by polaron excitation, while current conduction is caused by polaron hopping and the migration of oxygen vacancies at higher temperatures.

References

Bai, Y., Zhou, J., Gui, Z., Li, L., and Qiao, L. (2007). A ferromagnetic ferroelectric co-fired ceramic for hyperfrequency. *Journal of Applied Physics*, 101(8), 083907.

Cai, N., Shi, Z., Lin, Y., and Nan, C. W. (2004). Magnetic-dielectric properties of $NiFe_2O_4$/PZT particulate composites. *Journal of Physics D: Applied Physics*, 37, 823.

Dzunuzovic, A. S., Vijatovic Petrovic, M. M., Stojadinovic, B. S., Ilic N. I., Bobic, J. D. (2015). Multiferroic (NiZn) Fe_2O_4–$BaTiO_3$ composites prepared from nano powders by auto-combustion method. *Ceramics International*, 13189–13200.

Dzunuzovic, A. S., Vijatovic Petrovic, M. M., Stojadinovic, B. S., Ilic N. I, Bobic, J. D., Foschini, C. R., and Zaghete, M. A., and Stojanovic B. D. (2018). Magneto-electric properties of $xNi_{0.7}Zn_{0.3}Fe_2O_4$ – (1-x)$BaTiO_3$ multiferroic composites. *Ceramics International*, 44, 683–694.

Jian, X. D., Lu, B., Li, D. D, Yao, Y. B., Tao, T., Liang, B., Guo, J. H., Zeng, Y. J., Chen, J. L., and Lu, S. G. (2018). Direct Measurement of Large Electrocaloric Effect in $Ba(Zr_xTi_{1-x})O_3$ Ceramics. *ACS Applied Materials & Interfaces*, 10, 48014807.

Kumar, Y., Yadav, K. L. (2018). Synthesis and study of structural, dielectric, magnetic and magnetoelectric properties of $K_{0.5}Na_{0.5}NbO_3$–$CoMn_{0.2}Fe_{1.8}O_4$ composite. *Journal of Materials Science: Materials in Electronics*, 29(11), 8923–8936.

Ma, J., Hu J., Li, Z., and Nan, C. W. (2011). Recent Progress in Multiferroic Magnetoelectric Composites: from Bulk to Thin Film. *Advanced Materials*, 23, 1062–1087.

Mohan, S. and Joy P. A. (2019). Magnetic properties of sintered $CoFe_2O_4$–$BaTiO_3$ particulate magnetoelectric composite. *Ceramics International*, 45, 12307–12311.

Nan, C. W., Bichurin, M. I., Dong, S., Viehland, D., and Srinivasan, G. (2018). Mulriferroic magnetoelectric composites: historical perspective, status and future directions. *Journal of Applied Physics*, 3, 03110135.

Paul, J., Monaji, P., Reddy, V., Sakthivel J. K., Kumar, D., Subramanian, V., and Das, D. (2016). Synthesis, characterization, and magneto-electric properties of (1-*x*)

BCZT-*x*CFO ceramic particulate composites. *International Journal of Applied Ceramic Technology*, 12640.

Qian, X. S., Ye, H. J., Zhang, Y. T., Gu, H., Randall, C. A., and. Zhang Q. M. (2014). Giant Electrocaloric Response Over A Broad Temperature Range in Modified BaTiO3 Ceramic. *Advance Functional Material, 24*, 13001305.

Ramesh, T., Rajendar, V., and Murthy, S. R. (2017). $CoFe_2O_4$–$BaTiO_3$ multiferroic composites: role of ferrite and ferroelectric phases on the structural, magneto dielectric properties. *Journal of Materials Science: Materials in Electronics, 28*, 1177911788.

Rather, M. D., Samad, R., and Want, B. (2018). Improved magnetoelectric effect in ytterbium doped $BaTiO_3$ – $CoFe_2O_4$ particulate multiferroic composites. *Journal of Alloys and Compounds, 755*, 89–99.

Saffari, F., Kameli, P., Rahimi, M., Ahmadvand, H., and Salamati, H. (2015). Effects of Co-substitution on the structural and magnetic properties of $NiCo_xFe_{2-x}O_4$ ferrite nanoparticles. *Ceramics International*, 41(6), 73527358.

Sreenivasulu, G., Hari Babu, V., Markandeyulu, G., and Murty, B. S. (2009). Magnetoelectric effect of $(100 – x)BaTiO_3$–$(x)\ NiFe_{1.98}O_4$ (x = 20 – 80 wt %) particulate nanocomposites. *Applied Physics Letters, 94*, 112902.

Yamasaki, Y., Miyasaka, Kaneko S., He, J. P. Arima, T., and Tokura, Y. (2006). Magnetic Reversal of the Ferroelectric Polarization in a Multiferroic Spinel Oxide. PRL 96, 207204.

Zhai, J., Cai, N., Shi, Z., Lin, Y., and Nan, C. W. (2004). Magnetic-dielectric properties of $NiFe_2O_4$/PZT particulate composites. *Journal of Physics D: Applied Physics, 37*, 823.

Zhi, Y., Chen, A., Guo, R., Bhalla, A. S. (2002). Ferroelectric-relaxor behavior of $Ba(Ti_{0.7}Zr_{0.3})O_3$ ceramics. *Journal of Applied Physics, 92*, 2655–2657.

14 Sustainable computing – a new corridor with green computing

Nikita Jain[a], Gajendra Singh, Kamlesh Gautam and Abhishek Sharma[1]

[1]Associate Professor, Dept of Computer engineering, Poornima College of Engineering, Jaipur, Rajasthan, India

Abstract

Computers are now a common convenience in the modern human lifestyle. The need for energy in the IT sector has significantly expanded along with the rise in popularity of computing and IT services. Sustainable computing is a non-traditional kind of computing that aims to use computers and their peripherals efficiently, effectively, and in an environmentally friendly manner. Today's computers are so ingrained in our daily lives that existence would be meaningless without them. It is employed at households as well as at offices. The consumption of power generates a very high amount of carbon in our ecosystem as a result of the quick evolution of computing technologies. The best way to reduce this issue is to use as little power as possible to run computing equipment. This method is known as sustainable computing. Sustainable computing is a novel approach to "green computing," maintaining all environmental costs while sustaining our ecosystem with fewer energy emissions. The wide definition of sustainable computing aims to get the most out of technology while cutting down on wastes like power and emissions. This study focuses on green computing, its requirements, and the efforts that need to be taken in that direction. This essay also discusses how closely human existence is intertwined with current computers and computing equipment. No one can imagine living without a computer or other computing device, but the average person needs to be aware of how damaging these technologies are to the environment. In light of the rapidly developing IT technologies that are pervasive in our society, this article provides an overview of green computing methods.

Keywords: Green computing, Internet of Things, IT services

Introduction

Green computing is a strategy for ecologically responsible computer use. Using computing tools, processes, and resources in an environmentally friendly manner is known as "green computing." In order to reduce environmental dangers and pollution, computer modules and devices are developed, designed, engineered, manufactured, used, and disposed of businesses, organizations, governments, and people who use the technology are all involved in developing it. The topic of "green computing" is vast and involves several choices. at all levels, from big data centres enacting energy-saving regulations to people deciding not to use screen savers on their devices. Reducing the carbon footprints that IT systems leave is an efficient method. Growing industries that effect carbon emissions is made possible

[a]nikita.jain@poornima.org

DOI: 10.1201/9781003450917-14

by information and communication technologies (ICT). Green computing also employs management techniques and methods to cut down on energy waste

Literature review

This section provides a summary of the most important publications that have already been published in the literature that discuss the use of intelligent systems in green computing. Using Scopus from Elsevier, we discovered 255 papers that mention both fields. The significance and influence of both green computing and computational intelligence, a quick summary of some of the most pertinent publications is provided in this section. We only briefly summarise a sample of the 255 publications here; a scient metric analysis of the entire collection is provided in the section that follows. Nonetheless, in order to offer an impression of the types of research that have been done up to this point, we just briefly describe a few of the papers here. These summaries, in our opinion, inspire readers to look for additional information on these publications as well as in the other papers in these fields that are relevant to them.

The benefits of the suggested hybrid G-ANN technique were shown by the reported 51% energy reduction. The research presents an effective scheduling approach for green cloud computing that makes use of a 3D neural network predictor (Prakash et. al., 2019). One recent technology that is important in the digital age is cloud computing. The key issues in this situation are scheduling and load balancing. The strategy promotes green cloud computing by utilising less heat and power. In the work by, a new method for load balancing that combines hybrid intelligence and green computing is proposed (Carvin et. al., 2017). An adaptive neural fuzzy clustering method is used in the hybrid intelligent technique to load balance a collection of sensors. A new green cloud computing service built on fuzzy theoretical ideas is suggested in the study (Kashyap et. al., 2019b). Elasticity and economies of scale, two important characteristics of cloud computing, allow it to find numerous uses in the fields of science, business, and industry. In the paper (Ragmani et. al., 2018), a genetic algorithm and support vector model are used to propose a green computing allocation approach. By reducing performance based on a predefined fitness function, the genetic technique is employed to optimize the result. The proposed technique, when compared to the best-known algorithm, gets reduced energy usage and lowers violations with high throughput, according to simulation results (Singh and Mahajan, 2019). It has been demonstrated that the proposed A-GA strategy results in considerable energy savings (Kaur and Kaur, 2015). In the paper (Theja and Khadar Babu, 2015), a novel fuzzy logic method for the problem of choosing a green supplier is described. In the study (Mousavi et. al., 2020), a supply chain model for green computing is put out to solve industrial problems using a multi-objective approach to optimization. The parameters are considered to have a triangular fuzzy number shape because of the ambiguity of the data utilized in the model, which yields positive outcomes. The major objective of this research (Bera et. al., 2020) is to develop an intelligent strategy based on fuzzy logic theoretical notions for effective modelling of green composites machining (Bhowmik and Ray, 2019).

The goal of the research is to create the best working conditions for the synthesis of an ideal fuzzy system. To discover the optimal control strategy, fuzzy control systems of the types are constructed and compared (Kondratenko et. al., 2017). Today, it is common knowledge that traffic on metropolitan highways is growing daily. In addition, pre-timed traffic control is a common practise in cities nowadays, which is rarely a wise choice. This study offers a solution by utilising intelligent traffic flow prediction and control via green signalling, which yields excellent results. In order to accomplish adaptive behaviour in the allocation process, a reinforcement learning algorithm is used (Varghese et. al., 2013). A strategy for optimising green computing processes using a machine learning algorithm is provided in the work by (Karthiban and Raj, 2020). Data mining techniques are the best choice for resolving difficult issues in the healthcare sector because handling information is a crucial duty in this industry. The testing results are quite encouraging, and the technique may be useful for treating heart disorders, according to medical professionals. In the paper (Zubar and Balamurugan, 2020), a deep learning algorithm strategy for energy-aware dynamic resource management for attaining green computing. A brand-new energy-conscious resource management method is suggested in this study paper. The idea of combined virtual machines and the container consolidation strategy are used to accomplish this. Gholipour et. al. (2021) proposed approach is crucial and very significant in the field of green computing. In order to satisfy the paradigm of an energy-constrained heterogeneous Internet of Things (IoT) network, green computing has lately emerged as a possible alternative (Yang et. al., 2020). It is suggested that the objectives of green computing be accomplished using an architecture based on reinforcement learning. (Kashyap et. al., 2019a) This research makes use of an energy harvesting technique. Comparing the approach's simulation results to those of current conventional methodologies, they are good (Xu et. al., 2019). This significant influence could be quantified and measured in terms of the aims of green computing that will be attained more fully and effectively over time. Green computing is shown in Figure 14.1 given below:

Figure 14.1 Green computing

Green design is the process of creating digital products like computers, servers, printers, and projectors that are energy efficient.

- Green manufacturing: Reducing waste during the production of computers and other subsystems to lessen the impact of these operations on the environment.
- Green use: Making computers and their ancillary equipment use less electricity and in an environmentally responsible way.

Recycling unwanted electronic equipment or repurposing current equipment are both examples of environmentally friendly disposal.

With many opportunities to make it happen, green computing has a bright future for protecting the environment. hope we advance towards our objective of effective computing while making the earth greener. Maksimovic, (2017)

Parameters for power consumption

There are numerous measurement methods available to measure the electricity usage of data centers; some of them are listed here:

PUE, or power usage effectiveness PUE, or power usage effectiveness PUE is a calculation metric that measures the ratio of total energy consumption. It is used to calculate the energy efficiency of data centers. It's outlined as

$$PUE = \frac{TotalDataCentrePower}{ITDevicesPower}$$

Data center infrastructure efficiency (DCiE): It is the reciprocal of PUE; these tow electricity measurement techniques are very much popular between most of the data centers. It is defined as:

$$DCiE = \frac{1}{PUE}$$

$$= \frac{ITDevicesPower}{TotalDataCentrePower}$$

Compute power efficiency (CPE): It is a metric used to assess a data center's computing effectiveness as defined by CPE:

$$CPE = \frac{ITdevicesUtilization}{PUE}$$

$$CPE = \frac{ITDevicesUtilization * ITDevicesPower}{TotalDataCentrePower}$$

Green energy coefficient (GEC): The green energy consumption, or GEC, of a data center facility. Data centers are environmentally benign because all of these energies are produced using renewable energy sources. Definition of GEC:

$$GEC = \frac{GreenPower}{TotalFacilityPower}$$

Energy reuse factor (ERF): How much energy is reused from outside of the data center. It is defined as-

$$ERF = \frac{EnergyReused}{TotalFacilityPower}$$

Data center productivity (DCP): It is a dimension of the quantity of helpful work done by the data center. It is definite as:

$$DCP = \frac{ValuableWorkdone}{Totalresource}$$

Methodology

Green computing has been imagined advancing one of the most mind-blowing answers for issues of energy-utilization in processing. Green computing serves huge advantages to maintain conditions; in which energy-effectiveness positions as quite possibly. The following is the cycle wherein the strategy was directed: A converzation on the compromises was likewise viewed as which is profoundly critical in improving and featuring the qualities and shortcomings of every rules.

For this situation, a relative examination is profoundly vital for looking at and measure fluctuating strategies and methods of energy-effectiveness in green processing. This strategy for examination is especially reasonable for the goal of distinguishing the best energy-proficient procedure as it adds to a top to bottom comprehension of learned sources. On the other hand, an affirmation of the qualities and shortcomings of every model can undoubtedly be estimated through the examination of various methodologies.

The green computing implementation is shown in the Figure 14.2.

Preliminary results

The worth 'generally excellent' is expected for the review that exceptionally meets the prerequisites. Lastly, the worth 'medium' is assigned for the review that least meets the prerequisites, yet in addition conveys some basic substance. The fundamental reason for choosing these rules and naming them with a worth is because of the tremendous significance these elements incorporate. Finally, it carries out helpful supportability techniques by diminishing non degradable strategies, making it the most ideal choice contrasted with the rest. The second

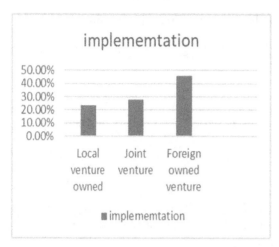

Figure 14.2 Green computing implementation

Table 14.1: Comparative study

S.NO	Author name	Year	Approach used	Findings
1	Alsamhi et. al.	2019	The paper presents a survey-based approach to examine the key approaches used include literature review, survey design and administration, data collection and analysis, and statistical modelling.	The survey was designed to capture the views and opinions of experts and researchers on the potential of IoT for green applications. Data was collected through an online survey and analyzed using descriptive and inferential statistics.
2	Bagla et. al.	2022	The paper examines a survey-based approach. The key approaches used include survey design and administration, data collection and analysis, and statistical modelling. The survey was designed to capture the knowledge, attitudes, and behaviors of IT professionals regarding green computing. Data was collected through online and in-person surveys and analyzed using descriptive and inferential statistics.	The paper finds that the most widely adopted green computing practices are virtualization, server consolidation, and energy-efficient hardware. Lack of management support, high costs, and lack of awareness were identified as the main barriers to adoption. The study highlights the need for more education and training on green computing, as well as the importance of government policies and incentives to promote adoption.
3	Castillo and Melin	2021	The paper presents a review-based approach to examine the key approaches used include literature review, data extraction and	The review was designed to identify and analyze the use of computational intelligence techniques in addressing environmental challenges and

S.NO	Author name	Year	Approach used	Findings
			synthesis, and analysis and discussion.	promoting sustainability. Data was extracted from peer-reviewed articles and analyzed using qualitative and quantitative methods.
4	Gholipour et. al.	2021	The paper proposes an energy-aware dynamic resource management technique for green computing in cloud data centers using a deep Q-learning algorithm and joint virtual machine (VM) and container consolidation approach.	The problem was formulated as a reinforcement learning problem, and the deep Q-learning algorithm was used to learn the optimal policy for resource allocation. Simulation was used to evaluate the performance of the proposed approach, and analysis was conducted to compare the results with other existing techniques.
5	Prakash et. al.	2019	The paper proposes a hybrid genetic artificial neural network (G-ANN) algorithm to optimize the energy component in a wireless mesh network (WMN) for green computing.	The findings suggest that the G-ANN algorithm can achieve better results in terms of energy efficiency and network performance compared to other techniques.
6	Kashyap et. al.	2019b	The paper proposes a neuro fuzzy logic-based load balancing technique for green computing in sensors-enabled IoT.	The findings suggest that the proposed approach can achieve better results in terms of energy efficiency and network performance compared to other load balancing techniques.
7	Ragmani et. al.	2018	The paper proposes the key approaches used include problem formulation, algorithm design, simulation, and analysis.	The findings suggest that the proposed fuzzy logic-based SLA can achieve better results in terms of energy efficiency and cost-effectiveness compared to other SLA negotiation techniques.
8	Bhowmik and Ray	2019	The authors used a set of input variables and output responses to build a soft computing model using adaptive neuro-fuzzy inference system and artificial neural network (ANN) techniques.	The findings suggest that the proposed soft computing approach can provide accurate predictions of the surface roughness quality, which can help improve the efficiency and effectiveness of green abrasive water jet machining.

hugest angle considered was the adjustments in the plans and assembling of gadgets. The comparative study is discussed in Table 14.1.

Conclusion

We have introduced a few creative thoughts for home clients as well with respect to cloud server farms to lessen power utilization and CO_2 discharge; we have likewise examined an alternate sort of force estimation procedures. Green processing is the need of current time, it gives harmless to the ecosystem registering power that trusts on energy proficient figuring, primarily it is centered around the decrease of CO_2 discharge to make IT industry contamination free. This study gives important data to academicians, research researchers and individuals inspired by scient metric insightful perspectives on green registering and computational knowledge methods. As future work, we envision the scientists advancing the use of summed up type-2 fluffy rationale, particular brain organizations, transformative and swarm knowledge strategies or crossover blends of these techniques, in green figuring for tackling genuine issues in numerous areas of use. It likewise means to urge flow scientists to re-examine past literary works and recognize holes and difficulties which might actually be stuck to in later proposition. A considerable lot of the articles are exceptionally predated and positively require space for additional exploration in the ongoing acts of Green IoT.

References

Alsamhi, S, Ma, O., Ansari, M., and Meng, Q. (2019). Greening internet of things for smart everythings with a green environment life: a survey and future prospects. *Telecommunication Systems*, 72(4), 609–632.

Bagla, R. K., Trivedi, P., and Bagga, T. (2022). Awareness and adoption of green computing in India. *Sustainable Computing: Informatics and Systems*, 35, 100745.

Bera, S., Jana, D. K., Basu, K., and Maiti, M. (2019). Novel Multi-objective Green Supply Chain Model with Emission Cost in Fuzzy Environment via Soft Computing Technique. *In International Conference on Information Technology and Applied Mathematics*, March, 1, 463–80. Cham: Springer International Publishing.

Bhowmik, S. and Ray, A. (2019). Prediction of surface roughness quality of green abrasive water jet machining: a soft computing approach. *Journal of Intelligent Manufacturing*, 30, 2965–79.

Carvin, L. B., Kumar, A. D. V., and Arockiam, L. (2017). ENNEGCC-3D energy efficient scheduling algorithm using 3-D neural network predictor for green cloud computing environment. *In International Conference on Intelligent Computing, Instrumentation and Control Technologies; Kannur(IN)*, (p. 1316).

Castillo, O. and Melin, P. (2021). Review on the interactions of green computing and computational intelligence techniques and their applications to real-world problems. *Journal of Smart Environments and Green Computing*, 1(2), 103–19. http://dx.doi.org/10.20517/jsegc.2021.01

Gholipour, N., Shoeibi, N., and Arianyan, E. (2021). An energy-aware dynamic resource management technique using deep q-learning algorithm and joint VM and container consolidation approach for green computing in cloud data centers. *Advances in Intelligent Systems and Computing*, 227–33.

Karthiban, K. and Raj, J. S. (2020). An efficient green computing fair resource allocation in cloud computing using modified deep reinforcement learning algorithm. *Soft Computing*, 24, 14933–42.

Kashyap, P. K., Kumar, S., and Jaiswal, A. (2019a). Deep learning based offloading scheme for IoT networks towards green computing. *In Proceedings - IEEE International Conference on Industrial Internet Cloud,* (p. 22).

Kashyap, P., Kumar, S., Dohare, U., Kumar, V., and Kharel, R. (2019b). Green computing in sensors-enabled internet of things: neuro fuzzy logic-based load balancing. *Electronics,* 8, 384.

Kaur, B. and Kaur, A. (2015). An efficient approach for green cloud computing using genetic algorithm. *In Proceedings on 2015 1st International Conference on next Generation Computing Technologies,* (p. 10).

Kondratenko, Y., Korobko, V., Korobko, O., Kondratenko, G., and Kozlov, O. (2017). Green-IT approach to design and optimization of thermoacoustic waste heat utilization plant based on soft computing. *Green IT Engineering: Components, Networks and Systems Implementation,* 1, 287–311.

Maksimovic, (2017). The role of green internet of things (G-IoT) and big data in making cities smarter safer and more sustainable. *International Journal of Computing and Digital Systems,* 6(4), 175–184.

Mousavi, S. M., Foroozesh, N., Zavadskas, E. K., and Antucheviciene, J. (2020). A new soft computing approach for green supplier selection problem with interval type-2 trapezoidal fuzzy statistical group decision and avoidance of information loss. *Soft Computing,* 24, 12313–27.

Prakash, B., Jayashri, S., and Karthik, T. S. (2019). A hybrid genetic artificial neural network (G-ANN) algorithm for optimization of energy component in a wireless mesh network toward green computing. *Soft Computing,* 23, 2789–98.

Ragmani, A., El Omri, A., Abhgour, N., Moussaid, K., and Rida, M. (2018). A novel green service level agreement for cloud computing using fuzzy logic. *In Closer 2018 - Proceedings of the 8th International Conference on Cloud Computing and Services Science,* (p. 658).

Singh, G. and Mahajan, M. (2019). A green computing supportive allocation scheme utilizing genetic algorithm and support vector machine. *The International Journal of Innovative Technology and Exploring Engineering (IJITEE),* 8, 760–766.

Theja, P. R., and Babu, S. K. (2015). An Adaptive Genetic Algorithm Based Robust Qo S Oriented Green Computing Scheme for VM Consolidation in Large Scale Cloud Infrastructures. *Indian Journal of Science and Technology,* 79175, 1–13.

Varghese, A., Bajaj, P., and Malik, L. (2013). Design of adaptive traffic flow control system with soft computing tools for green signaling. *In 2013 International Conference on Human Computer Interactions.*

Xu, L., Qin, M., Yang, Q., and Kwak, K. (2019). Deep reinforcement learning for dynamic access control with battery prediction for mobile-edge computing in green IoT networks. *In 2019 11th International Conference on Wireless Communications and Signal Processing.*

Yang, M., Yu, P., Wang, Y., Huang, X., Miu, W., Yu, P., Li, W., Yang, R., Tao, M. and Shi, L., (2020). Deep reinforcement learning based green resource allocation mechanism in edge computing driven power internet of things. *In 2020 International Wireless Communications and Mobile Computing (IWCMC),* (pp. 388–393). IEEE.

Zubar, A. H. and Balamurugan, R. (2020). Green computing process and its optimization using machine learning algorithm in healthcare sector. *Mobile Networks and Applications,* 25, 1307–18.

15 Cost effective contemporary developments in trickle impregnation technology

Mahadev Gaikwad[a] and Rajin M. Linus

Department of Electrical and Electronics Engineering, Sanjay Ghodawat University, Kolhapur, India

Abstract

This article highlights the benefits of contemporary developments in rotating table type trickle impregnating machines with less investment rather than go for new one with huge capital investment. This contemporary development helps in wastage reduction, enhancement of productivity, maintenance cost dropdown to zero, environment friendly process, enhancement of operator's health, avoid fatigue, improve product quality and deduct the process cost to optimum with the help of programmable logic control (PLC) logic. Also, it eliminates old relay contactor logic, issue of current and temperature control, Induction motor dispensing system which doesn't have control on impregnating resin.

Keywords: Trickle impregnation, contemporary, cost saving, productivity, programmable logic control

Introduction

Nowadays various research activities have been in progress in the area of electrical machine as those machines are in demand due to the developments in electric vehicle and renewable energy especially wind energy conversion system. However, contemporary developments in electrical machines elevate the attractiveness of the manufactures toward the insulation of armatures and fields. The relay type contactor system with limited automation shows the ineffectiveness due to the continuous maintenance and non-availability of spares. In today's scenario, there are lot of options available in impregnation machines (Academic, 2022) and machine manufacturers made available the programmable logic control (PLC) operated machines with customer proposed features. Following are the details of different features of impregnation technologies with high quality performance. The impregnation techniques in shows that there will be no after work process as no need to remove trickles in the commutator slots, outer diameter of lamination, insulation paper and shaft, maximum resin filling on the winding slots, very high bond-strength standards and excellent resin penetration in the whole winding (Academic, 2022). They have short process times and high productivity. However, some of the shortcomings of those methods, includes these machines are not suitable for small or medium scale industries whose production rate is less. The conveyor index mechanism may face the issues such as less index accuracy and chain drop in the machine. The wear and tear problem

[a]gaikwadmbg@gmail.com

DOI: 10.1201/9781003450917-15

in the system is ignorable. Also, the space requirement of the system is more as the panel and machine are not integrated together. Apart from that the capital cost is too high.

Summaries to above merits and demerits, this project assures and leads toward cost effective contemporary developments in trickle impregnation technology. These recent enrichments have been possible with the use of synchronization of PLC and sensors. This proposed automation has eliminated old bulky relay contactor logic which results maintenance and repair cost get down to zero. The close monitoring and controlling of temperature through remote infra-red temperature sensor helped in online inspection and sorting of faulty components instead of generating waste. Precise temperature control with preheat-trickling-post curing process in this proposed system assures end quality output. This proposed impregnation process implemented in induction motor and dc series motor with peristaltic pumps dispenses gel coat with PLC interfacing. Hence, the manual efforts and intervention have been nullified. Also, it assures optimum Trickle quantity penetration, bonding of coil, zero bubbles trapping, no air cavity and improved accuracy of indexing.

The standard trickle impregnation is a method of coating the field and armature which is clamped on a mandrel and continuously rotated around its own axis in horizontal alignment (Cincinnati et al. 1985).

The process involves preheating the windings, heating during dispensing and post curing of the resin. The component is heated by conventional furnace heating or resistive/resistance/Ohmic heating. During rotation, the low-viscosity resin is trickled through several nozzles at different positions by means of peristaltic pumping system based on moving a product through a hose, by compressing and decompressing. The resin penetrates the windings and is distributed evenly in the armature and field which results high quality resin impregnation. These machines have been designed with the dial or "rotating table" in either the vertical or horizontal plane (Figure 15.1)

Dial type trickling machines are modern devices for the Volumatic impregnation of armatures and fields using the trickling method.

The multi-cycle trickling process for armatures and Fields are used here. This method not only has considerable advantages over the conventional dipping

Figure 15.1 Dial type impregnation machine
(*Source:* M/s Eibenstock Positron, India)

method, but it also means a great improvement over the well-known and normally used single-stroke trickle method.

In particular, the impregnation times are significantly reduced. The windings are thoroughly impregnated on all sides and have excellent electrical and mechanical properties after trickling.

Completely wound and switched stators or armatures are automatically preheated, trickled and cured on the system according to an adjustable program as shown in Figure 15.2. The heating takes place under rotation by means of a controllable flow of current through the coils, with the current being fed to the cable outlets of the station or the cable via slip rings and terminals. is fed to the collector. The dressing takes place at six stations before and five stations after the trickle. Also, with read the trickling itself (up to five stations) can be heated.

Each heating zone has a separate regulating transformer. This allows clean temperature control throughout the entire process.

The objects are dripped at 24 consecutive stations according to the multi-cycle process specially developed for this machine. To make it shorter refer armature field processing sequence given in flow chart in Figure 15.3.

Armature and fields are rotated by chain drive which rotates the spindles mounted on rotating/indexing table at an appropriate speed suitable for trickling. Rotating spindles are mounted on table driven by asynchronous motor used rotary index table. Controller with proportionate temperature control. Gelling is fed by feedback controlled peristaltic pump. Entire cycle is automatic excluding loading and unloading. The article flow is as follows: Introduction is in section I, section II consists of proposed system, discussion is in section III, conclusion IV and the acknowledge is in section V.

Literature review

The impregnation techniques have been installed since the year 1980. These machines installed with comparatively bigger size panels as its complicated electrical wiring and a greater number of relays required. PLC can be considered as

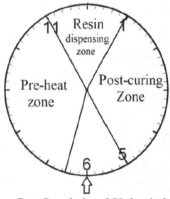

Part Loaded and Unloaded

Figure 15.2 Operational details of rotating table type set up

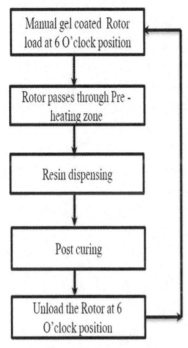

Figure 15.3 Existing armature and field processing sequence

Figure 15.4 Proposed Impregnation Machine
(*Source:* M/s Eibenstock Positron, India)

a compact box which controls thousands of digital relays and ability to handle complex tasks (Academic, 2022).

Figure 15.5 elaborates the process flow improvements and added quality checks after contemporary developments in existing machine. There are two

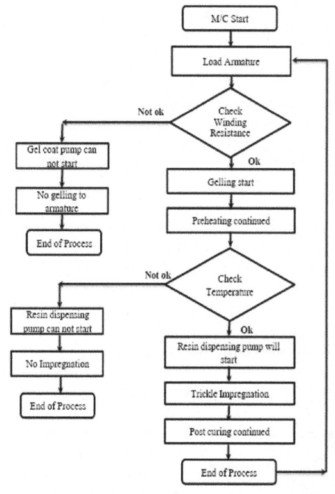

Figure 15.5 Proposed system armature and field impregnation flow chart

major controls are added to ensure the pre requirement and the further process will be continued. The first control confirms the resistance of part under gelling. If its resistance value has found within specification, then gelling pump starts and does gel coating at commutator side of armature otherwise no. Another control works on same working principal only change is here temperature get measure through non-contact type of sensor instead of resistance. If temperature of part under test has reached to its specified temperature band trickling pump starts. Both the controllers are helps to arrests faults in the wound armature, its end result is zero rejection.

1. Gelling

Manual gelling atomized and new station made at the seven o'clock position. Here gelling will be dispensed after getting confirmation of winding continuity.

Figure 15.6 emphases the gelling process. Gelling dispenses became more effective, appropriate amount of gel coat penetrates in the connection cavity of coil and commutator. Elimination of manual intervention has reduced the rejection and improved the final product.

Hence, process parameter setting for examples indexing, gelling, impregnation time, input voltage for armature heating became easy due to all provided at one place through operating switches and measuring Instruments as in Figure 15.7.

2. Resin dispensing

At resin (Academic, 2022) dispensing temperature is monitored through not contact type temperature indicator (Figure 15.8) if the temperature reached to set temperature indicator will become green and the peristaltic pumps start to dispense the resin. It dispenses set quantity. To avoid wastage as well as last drop fall on lamination peristaltic pump program and wiring modified to change its direction reverse.

Advanced infra-red temperature sensor measures temperature of moving parts with precise accuracy. Very short measurement and response times further control became more accurate.

Figure 15.6 Gel coating through pump
(*Source:* M/s Eibenstock Positron, India)

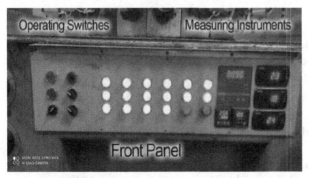

Figure 15.7 Front panel
(*Source:* M/s Eibenstock Positron, India)

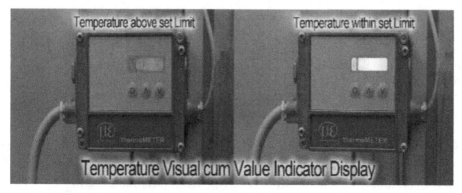

Figure 15.8 Non-contact type temperature indicator
(*Source:* M/s Eibenstock Positron, India)

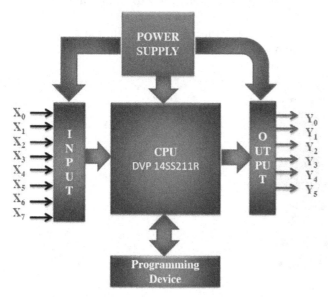

Figure 15.9 Block diagram of PLC input and output for model DVP14SS211R

3. Indexing mechanism

Asynchronous motor used rotary Index table (Academic, 2022). Extremely long service life combined with impressively fast switching. Robust rotary indexing table with smooth, jerk and impact-free running and extremely long service life.

4. PLC

The resultant output has been achieved with the help of both PLC's DVP14SS211R and DVP16SP11R (Academic, 2022). The programming input and output details are explained through block diagram (Figures 15.9 and 15.10).

There are two I/P – O/P and feedback for Trickle Impregnation Process-DVP 14SS211R (Figure 15.5).

Input
  X_0 – Start push button
  X_1 – Stop push button – Auto/manual selector switch
  X_2 – Trickle 1 manual on selector switch
  X_3 – Trickle 2 manual on selector switch
  X_4 – Gel coat manual on selector switch
  X_5 – Cycle time input
  X_6 – Gel coat dispensing time

Output
  Y_0 – Main motor on (armature rotating motor)
  Y_1 – First station motor on
  Y_2 – Main contactor on
  Y_3 – Sol valve 1 on (index lock/unlock sol)
  Y_4 – Sol valve 2 on (index on)
  Y_5 – Sol valve 3 on (index on)

DVP 16SP11R (Academic 2022)

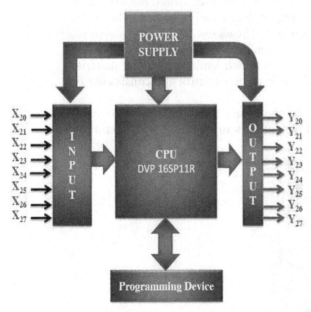

Figure 15.10 Block diagram of PLC input and output for model DVP16SP11R

Input
  X_{20} – Temperature input
  X_{21} – Trickle 1 CPR I/P
  X_{22} – LS 1 trickle station home
  X_{23} – Trickle 2 CPR I/P
  X_{24} – Table index 1 proxy
  X_{25} – Table index 2 proxy

X_{26} – Hydraulic pressure LS
X_{27} – Gel coat CPR input

Output
  Y_{20} – Trickle 1 forward start
  Y_{21} – Trickle 1 reverse start
  Y_{22} – Trickle 2 forward start
  Y_{23} – Trickle 2 reverse start
  Y_{24} – Gel coat forward start
  Y_{25} – Gel coat reverse start
  Y_{26} – Gel coat timer reset
  Y_{27} – Gel coat timer reset

Peristaltic pump with PLC configuration and controlling system shows extremely effective result as filling level increased, less wastage and provide better mechanical strength. The impregnation is also called secondary insulation because it reinforces the copper enamel which is defined as the first insulation.

Non-contact type temperature measurement systems in Figure 8 with short response time have avoided rejection of part under impregnation. Robust rotary indexing table with smooth, jerk and impact-free running has dropdown the indexing mechanism maintenance cost to zero (Academic, 2022).

Reducing environmental footprint, streamlining the processes within line quality increasing quality of process output which leads to minimal material wastage of operation. Ultimately results into greater control and consistency of product quality.

Result of improvement is witnessed by the Figure 15.11. Before development lot of Armatures and Fields rejected for bubbles on winding, no penetration

Figure 15.11 Comparison of output before or after contemporary development
(*Source:* M/s Eibenstock Positron, India)

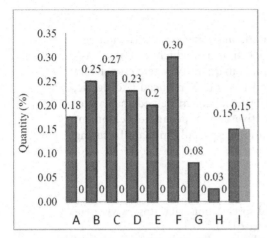

Figure 15.12 Comparison of armature rejection summary: A- trickle spread on stack, B - trickle spread on wedge insulation C - trickle enters in commutator slot, D- cavity in winding, E - wire loose, F- bubbles on winding, G- poor aesthetic, H - gelcoat spread unevenly, I- power failure

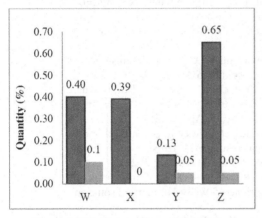

Figure 15.13 Comparison of field rejection summary: W-wire loose, X-bubbles on winding, Y- resin spread on lamination inner diameter. Z- resin on insulation paper

of resin in winding cavity and aesthetic. The proposed system solved all issues and approximately 15 - 18 lac saved due to armature and field Rejection. Also, approximately there was 2 lac rework cost waver per year.

Figure 15.12 explains the armature rejection % comparison before and after contemporary development. Alphabets (A, B, C......I) explain different types of rejection exist in Armature Impregnation. In this entire study except in the rejection during power failure, all rejections are zero.

Figure 15.13 explains the field rejection % comparison before and after contemporary development. Alphabets (W, X, Y, Z) explain different types of rejection exist in Armature Impregnation. When compared to previous rejection, there is a drastic and considerable reduction which is varying between 0.05 and 0.1%.

Conclusion

The development of rotating table type automatic machine is the correct choice it has saved approximately 1720 lac per year rejection cost which is ¼ of project cost. Excellent quality and countable improvement in quantity is delivered. Resin wastage and part under impregnation rejection has totally reduced (Academic, 2022 and Altana I. P. D., 1985). New Indexing mechanism helped in maintenance cost dropdown to zero. This cost-effective contemporary development solution is very successful in industrial applications especially in manufacturing industries.

References

Academic trickle-varnishing-machine. Features and specification data. ttps://www.aren-ginearengineeringindia.com/. (Accessed on 4 Aug 2022).

Academic impregnation methods, machines details. https://www.bdtronic.com/en-en/impregnation-methods/trickling/. (Accessed on 4 Aug 2022).

Academic trickle coat systems available around the world. https://www.heattek.com/trickle-coat-systems.html. (Accessed on 4 Aug 2022).

Academic trickle impregnation machines. https://www.stator-systems.com/index.php?page=shop.browse&category_id=8&option=com_virtuemart&Itemid=63. (Accessed on 5 Aug 2022).

Academic impregnation machine latest features and its prize. https://www.motor-machinery.com/Armature-production.html. (Accessed on 5 Aug 2022).

Delta Programmable Logic Controller DVP Series Academic PLC product details referred for the selection. https://www.deltaww.com (Accessed on 5 Aug 2022)

Adamic Delta Electronics, Inc. Taoyuan Technology Center No.18, Xinglong Rd., Taoyuan City,). DVP-ES2/EX2/SS2/ SA2/SX2/SE&TP Operation Manual – Programming (2018) https://www.deltaww.com/en-US/index. [Date. 04 Apr 2022]

Academic electrical insulation and its properties. https://www.elantas.com/beck-india/products/electrical-insulation-system/electrical-insulating-varnishes-resins.html. Product selection Guide - Impregnating Resins. (Accessed on 4 Aug 2022).

Academic rotary index table technical details. https://www.weiss-world.com/in-en/products/rotary-indexing-tables-44/rotary-indexing-table-45 Weiss Automation solutions India Pvt. Ltd. (Accessed on 12 Aug 2022).

Speer, D. R., Sarjeant, W. J., Zirnheld, J., Gill, H., & Burke, K. (n.d.). Insights into coil processing. Proceedings: Electrical Insulation Conference and Electrical Manufacturing and Coil Winding Conference (Cat. No.01CH37264). DOI : 10.1109/eeic.2001.965740

Timmins, P. A., & Morgan, J. (1985). Epoxy "trickle" impregnation with solventless resin vs. dip and bake in solvent varnish. 1985 EIC 17th Electrical/Electronics Insulation Conference. doi:10.1109/eic.1985.7458643

16 Analysis of rotary over traffic signal for sustainable urban development

Praveen Bhardwaj[a] and Dr. Piyusha Somvanshi

Poornima College of Engineering, Jaipur, India

Abstract

In this paper we analyze rotary vs traffic signal system to achieve goals of urban sustainable development. This study is done by taking different parameters i.e., journey time, reduction in the distance travelled, and delay time. We looked at the locations in the city where five different roads converge. We selected Sheikh Zayed rd for our comparative study. By using a video graphic survey method, we gathered traffic data during the busiest times of the day. We then divided the vehicles into various classes. Upon approaching the rotary, cars and trucks are obligated to slow down but do not stop as they would at a signalized intersection. This could shorten the time that cars spend waiting at intersections. The rotary's only expense is its construction as ongoing upkeep is quite cheap. PTV VISSIM (Planung Transport Verkehr Verkehr In Städten - SIMulationsmodell (German for "Traffic in cities - simulation model")), a simulation tool, is used for design purposes. We used this software to construct the signalized intersection and roundabout at the Sheikh Zayed rd intersection. Design the signalized intersection and rotary by depicting the linkages, connectors, indicating the reduced speed area, inserting the nodes, and determining the appropriate cycle length in this manner by adhering to the correct design procedures. On the basis of the signal, the travel and delay times are determined. The rotary intersection's characteristics are derived using the same vehicle data that was entered into the Rotarystock returns.

Keywords: Dubai, intersection, signalized intersection, traffic, rotary

Introduction

The traffic rotary, sometimes referred to as the rotary junction of roads, is just a long intersection of roads where vehicles can pass or turn without stopping. All the vehicles travelling in different directions circulate around the central island in a single direction before diverging at the necessary exit.

A shared place that is utilized by many approaches simultaneously is an intersection. When an intersection is signalized, a set number of approaches use the shared space in alternating fashion for a predetermined period of time in accordance with the intersection's phasing system. The primary aspects of such an intersection that aid in its analysis and design for a specific level of service are its delay and queuing mechanism.

The amount of traffic in Dubai has significantly increased during the past ten years. As the number of vehicles increases, congestion rises, and the area becomes more congested. There is consequently a delay in getting there. The three types of total delay are acceleration delay, stop delay, and declaration delay.

[a]Piyusha.somvanshi@poornima.org

DOI: 10.1201/9781003450917-16

In general, traffic signals are crucial to the efficient operation of the urban street traffic system.

A good signal aids in enhancing road mobility and easing traffic in both rural and urban areas. Due to the traffic's rapid and unchecked growth, Dubai is experiencing traffic congestion at many road intersections. Transportation system that causes environmental deterioration, traffic delays, and fuel waste. In most cases, drivers cannot turn off their engines while waiting for their turn to cross a junction and instead honk or blow their horns unnecessarily, which delays the vehicle, uses more fuel, and creates noise pollution at all signalized intersections.

In a case study project called "Comparative study on signalized intersection and rotary," due to the rapid and unchecked growth in traffic, the traffic situation in a particular area of Dubai is evaluated at a certain intersection with multiple roads. Transportation system that causes environmental deterioration, traffic delays, and fuel waste. In most cases, drivers cannot turn off their engines while waiting for their turn to cross a junction and instead honk or blow their horns unnecessarily, which delays the vehicle, uses more fuel, and creates noise pollution at all signalized intersections.

In a case study project called "Comparative study on signalized intersection and rotary," the traffic condition in a particular area of Dubai is assessed at a location where the roads have five intersections. The study is then conducted using a videography method to count the number of vehicles on a square. After analyzing the collected data, the number of two-wheelers, three-wheelers, four-wheelers, and heavy vehicles during the busiest hours of the day is then calculated. The delay at the signalized intersection is then estimated using the VISSIM software, and it is contrasted with the delay at the rotary.

Literature review

Fromme (2010) made an effort to research how traffic platoons behave at intersections in the city of Al Bastakiya. Using a videographic survey technique, we can measure how much is the traffic volume. They examined the recommended route to calculate the peak traffic flow. They performed three experimental programmes: rotary or roundabout intersections, traffic signal designs, and vehicle volume counts. Due to the volume of traffic, however, the attempt to create the signal was unsuccessful, and it was suggested for a rotary.

Shirazi et. al. (2022) investigated traffic and discovered that right-turning vehicles and those coming from the other end create the majority of confrontations at intersections. The prerequisites for establishing a left turn waiting area under a special right turn face are examined.

The main focus of Chauhan et. al. (2020) research on the new concept of traffic rotary design and road intersections was on right turn moments. In addition, they discussed how to explain rotary design, design speed, entry-exit Island radius, and the geometry of the Central Island, as well as how to calculate rotary capacity using an empirical method.

Otto and Simeon (2022) studied three parameters, including volume, speed, and capacity because these parameters have a significant impact on traffic laws. They also collected data to calculate the length of the road and the level of service, and speed data was calculated at the middle of a six-lane road in Dubai Marina.

On the first day of their assessment of the traffic situation, they suggested the design of a fixed time signal in place of a Rotary intersection in Al Karama, Saha and Sobhan (2012) tried to deal with the problems of traffic clogging up and extraordinary delays in moving traffic. They then converted the value into PCU. From their investigation, they concluded that the traffic approaching at the crossroads is quite high, 3000 PCU per hour. Seven days of data were collected in a week of various time intervals from morning to evening peak hour.

Methodology

```
┌─────────────────────────────────┐
│        Section of a site        │
└─────────────────────────────────┘
                 │
                 ▼
┌─────────────────────────────────┐
│  Data collection method and     │
│  analysis via a video survey    │
└─────────────────────────────────┘
                 │
                 ▼
┌─────────────────────────────────┐
│  Calculation of various         │
│  quantities using various       │
│  formulae                       │
└─────────────────────────────────┘
                 │
                 ▼
┌─────────────────────────────────┐
│  Analysis of various parameters │
│  in rotary and signalized       │
│  intersection                   │
└─────────────────────────────────┘
                 │
                 ▼
┌─────────────────────────────────┐
│  Comparison of rotary and       │
│  signalized intersection        │
└─────────────────────────────────┘
```

Site selection

For collection of data related to traffic, we must select some places that have rotary and signalized intersections. In our case we have selected some areas near Sheikh Zayed rd, Al Bastakiya city, Dubai Marina and Al Karama.

Data collection and analysis using videographic survey

After selection of site, we decided to select the videographic method for our data collection. In which we shoot many videos of moving traffic as well as note the various parameters such as time, speed, delay, traffic density, etc.

Calculation of various quantities using various formulae

After collection of all the required data, we calculated various parameters such as speed, traffic flow, traffic density, spot speed, time mean speed, space mean speed using various formulae.

Analysis of various parameters in rotary and signalized intersection

After calculation of all the data, we analyze the data of both rotary as well as signalized intersection of all the places that we have selected.

Comparison of rotary and signalized intersection

The last step is to compare the data of rotary & signalized intersection and conclude the results.

Results and discussion:

The following tables shows the observed data of Q-length and Vehicle delay of four different locations of Dubai i.e. Sheikh Zayed Rd, Al Bastakiya city, Dubai Marina and Al Karama.

Table 16.1: Day 1 morning (signalized intersection).

	Sheikh Zayed Rd	Al Bastakiya city	Dubai Marina	Al Karama
Q- length	100.02	100.02	100.02	100.02
Vehicle delay	161.10	184.26	167.02	149.03

(*Source:* Author's compilation)

Table 16.2: Day 1 morning (rotary).

	Sheikh Zayed Rd	Al Bastakiya city	Dubai Marina	Al Karama
Q-length	98.12	98.12	98.12	98.12
Vehicle delay	100.55	168.45	82.26	82.13

(*Source:* Author's compilation)

Table 16.3: Day 1 evening (signalized intersection).

	Sheikh Zayed Rd	Al Bastakiya city	Dubai Marina	Al Karama
Q-length	96.01	96.01	96.01	96.01
Vehicle delay	150.40	160.82	164.43	159.98

(*Source:* Author's compilation)

Table 16.4: Day 1 evening (rotary).

	Sheikh Zayed Rd	Al Bastakiya city	Dubai Marna	Al Karama
Q-length	82.86	82.86	82.86	82.86
Vehicle delay	82.55	50.26	72.16	28.03

(*Source:* Author's compilation)

Table 16.5: Day 2 morning (signalized intersection).

	Sheikh Zayed Rd	Al Bastakiya city	Dubai Marina	Al Karama
Q-length	95.45	97.23	90.80	93.76
Vehicle delay	140.43	139.76	146.88	145.22

(*Source:* Author's compilation)

Table 16.6: Day 2 morning (rotary).

	Sheikh Zayed Rd	Al Bastakiya city	Dubai Marina	Al Karama
Q-length	59.45	59.45	60.78	57.32
Vehicle delay	160.32	135.78	156.68	175.45

(*Source:* Author's compilation)

Conclusion

The present review is conducted for analyzing the comparison of rotary over traffic signal in various areas of Dubai. In order to be suitable for traffic flow with greater traffic volume from each intersection of the square, this evaluation also analyses the wide range of control delay and fuel consumption. The design of rotaries at road intersections is examined using a novel geometric idea. It has many uses in the planning and design of road improvements. The study focuses on crucial variables that affect collisions and delay.

References

Chauhan, G. S., Musa, A. A., and Garba, L. (2020). Performance assessment of rotary intersection capacity (case study of baban gwari roundabout), Complience Enginering Journal, ISSN no 0898-3577, 207-213.

Fromme, V. (2010). Research on Roundabouts, Report Written by Victoria Fromme Winter 2010 Neihoff Studio.

Otto, C. G. and Simeon, B.(2022). Capacity assessment of slaughter rotary intersection. Journal of Newviews in Engineeringand Technology, 4(1), April 2022, 14-21.

Saha, A. K. and Sobhan, M. A. (2012). Features & facilities at C&B road intersection: a case study.IJASETR, ISSN: 1839-7239, August 2012, 1(4), 19-28.

Shirazi, M. S., Chang, H. F., and Tayeb, S.(2022). Turning movement count data integra tion methods for intersection analysis and traffic signal design, Sensors, MDPI, 2022, 22(19), 7111, https://doi.org/10.3390/s22197111.

17 Study in India's charging infrastructure

Ashish Mandal[a] and Himani Goyal Sharma[b]

Department of Electrical Engineering, Chandigarh University, Punjab, India

Abstract

The lack of petroleum and diesel and fossil fuel expanding costs of fuels, and an aggressive program by the Indian government has empowered multidisciplinary research on electric vehicles& assortment of EV charging stations as an option to customer ICE vehicles. The charging framework is the centre necessity of electric vehicle market development. As of now situation, the absence of a charging framework and higher battery costs stay a central issue region for mass transformation of EV. This paper talks about how EV charging foundation is one of the significant obstacle for EV transformation in India. In this setting, examination here clarifies the elements influencing Variation, the accessible charging foundation in India, charging principles, challenges in the development of the charging framework and improvement needs are plainly clarified. The continuous Modern city attempt in an India needs to deal with issues such as air pollution contamination, expanding ozone-depleting substance discharges, and rising dangers of energy security. In numerous urban communities the levels of carbon, Sulphur, and nitrogen oxides in India's urban air pollution are astronomically high.

Keywords: CHEDMO charging, level 1 charger, Level 2 charger, fast Speed charging, EVSE

Introduction

Electric vehicle Infrastructure technicians are in high demand right now in bright future and career in Installation, operation, and maintenance of charging Infrastructure for Electric Vehicles it is very interesting work for bright future. Today we have facing a lot of problems lacking of electric vehicle charging infrastructure Chan, C.C (2007). Now people are aware of Electric vehicle some company is taking very good step as they are provide home vehicle charger to charge a vehicle. It is very necessary to our country because nowadays day now a days increasing pollution by internal combustion it is very harmful impact to our health Foley AM (2010). In my point of view. The best alternatives to replace conventional cars, in my opinion, are electric and hybrid vehicles. However widespread adoption of electric mobility and mobility will necessitate residential and public charging infrastructure and a switch from traditional fuel pumps. Many countries throughout the world have established level 1, level 2, and rapid electric vehicle speed charging stations. vollers (2013).It is necessary to conduct research in order to establish a suitable infrastructure for the Indian market.

Charging levels

Table 17.1: Charging Levels with its min and max rating.

[a]ashi999999@gmail.com, [b]himani.e10806@cumail.in

DOI: 10.1201/9781003450917-17

Table 17.1: Charging levels with its min and max rating.

Level of charging	Minimum power rating (kw)	Maximum current rating (A)
AC Charging		
Level-1	47.5	16
Level-2	8.15	32
Level-3	60120	250
Fast DC charging	100200	400
	SAE standard	
AC charging		
Level-1	2	16
Level-2	20	200
Level-3	Above 20	
DC charging		
Level-1	2	80
Level-2	20	200
Level-3	Above	400
	CHAdeMO	
DC fast charging	62.5	125

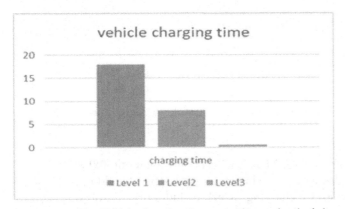

Figure 17.1 Vehicle charging time according to level of chargers (Mayfield, 2016)

Level-1: In figure 17.1 charging equipment's provides charging through a normal 120 AC plug. Level 1 One type of charging apparatus is portable and does not require installation. You just need to insert the socket and we can charge your vehicle. Depending on min and max rating.

Level-2: In figure 17.2 charging equipment's offer charging to a 240V in AC plug and this type of charging level required installation of public and residential charging infrastructure these units are required the dedicated circuit special charging equipment's u need to install this to charge the vehicle. So in The all-electric vehicles can use level 2 charging equipment and depending upon the

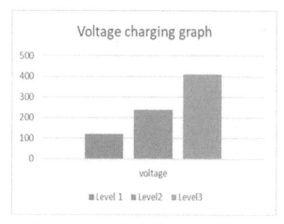

Figure 17.2 Consume voltage according to level of chargers (Karkatsoulis et. al., 2017)

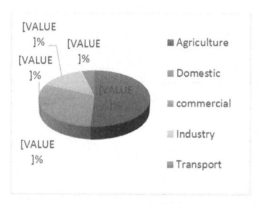

Figure 17.3 Sectors with the highest atmospheric carbon dioxide emissions (Karkatsoulis et. al., 2017)

battery technology used in the vehicle level 2 charging generally takes 06to08hrs to completely charge fully degraded battery see in fig2 Only residential buildings and public parking spaces have level 2 charging, which includes organizations like your offices.3. Level 3 charging or DC fast with this charging method, the on-board battery charger for electric vehicles is bypassed, and direct current is used to transmit energy. The majority of level 3 chargers deliver 80% of a charge in 20- 30 Minutes.

The charging infrastructure for electric vehicles

Electric vehicle charger is a very important this is a component of the charging infrastructure. Time to charge a battery electric vehicle has totally dependent on type of charger. Good charger is should have a high power density and be light in weight. An Electric vehicle chargers must insured a high-quality utility current with minimal distortion so it is iminium Grid electricity quality has an impact..

Insured low disturbance is produced within the current's source the power factor drops as a result of this. The government is giving a better subsidy to establishing in electric vehicle charging system. In figure 17.3 showing different types of sectors are consuming the pollution as per the graph.

Types of electric vehicle charging

AC and DC charging: The AC charger is the most common charging method for electric vehicle with the plugs. When we a plugin normal point Electric vehicle is a normal charging point the power gets converted in AC to DC inside the on-board charger. The on-board charging is doing it is converting ac power is coming from the ac charger top dc which will move to the batteries'. Jahan S(2015) So this is doing convert in to ac to dc and ac charger 3.3 kW to 22 kW in single phase and three phase respectively. The ac charger can basic but it can be smart with Wi-Fi connectivity, LAN connectivity etc. based on EVSE controller inside the AC charger. EVSE stand for electric vehicle supply equipment. Controller which will decide your charger is smart or basic type. The electricity goes directly into the car battery, without the converter in the vehicle. Ilieva, Liliya (2016) Hence these are fast charger and dc charger ranges from 30kw to 350kw with high rating in used for charging heavy vehicles such as bus, truck etc.

Plug or types of connector

In figure 17.6 now, we have different types of Ac charging and DC charging

AC connector

Type 1 plug 3.7to 7kw single phase with 5 pins and it has no locking mechanism.
    Type 2 plug 3.7 to 7 kW single phase and up to 22kw three phase with 7 pins and it has an inbuilt locking mechanism. It is also inbuilt automatic locking system. They are two plug use in public charging area.

DC connector

Charge de Move, which is short for CHAdeMO and is equivalent to the phrase "charge n go," is a fast charger. It was proposed as a plan in 2010. Five Japanese automakers formed the same-named organization. Which has established a world standard for the sector. Integrated charging system, or CCS. For the purpose of connecting a high voltage DC charging station to the vehicle's battery, two additional connectors are added to the bottom of type 1 or type 2 vehicle inlets and charging plugs. Somayaji Y (2017) These connections are frequently referred to as combo 1 or combo 2.GB/T it is DC charging standard used in china.

Charging infrastructure in electric vehicle

Use Today we have electric vehicle market extremely depending on the vehicle the availability of charging infrastructure. So the developing infrastructure for electric vehicle of charging is most important Governments have a lot of work

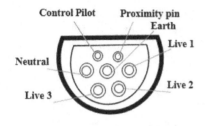

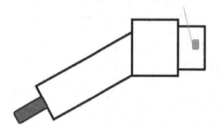

Figure 17.4 AC connector type 2 (Chen et. al., 2020)

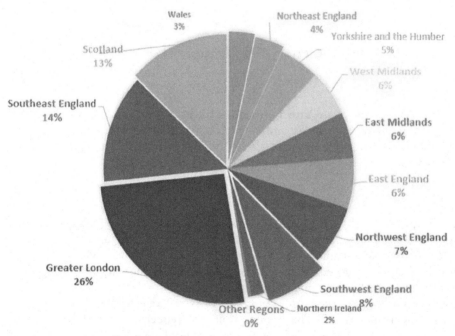

Figure 17.5 Profile of charging connector (Chen et. al., 2020)

ahead of them, and one of them is electric automobiles. Manufactures. It is very important factor is the exciting power must be bear the accommodate the charging load. Vollers (2013) Availability of commercial charging system is make very helpful to us. These are benefits of the power grid.

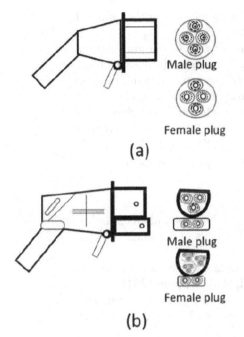

Male plug

Female plug

(a)

Male plug

Female plug

(b)

Figure 17.6 DC connector. (a) CHAdeMO. (b) Combo 2 or CCS (Chen et. al., 2020)

- It will assist in diverting the charging load of the network's pick demand.
- When the bulk charging load is increase of harmonics and reduce the power factor.
- Implementation of vehicle to grid system would be easy reduce the need of communication, and control up to the consumer. D. Mayfield (2016).
- The installation of the electric car charging station should be chosen such that it meets the requirements of both the consumer and the electrical grids. Sarker MR (2017)

Charging modes

The international; Electro technical commission defines charging in modes (IEC62196).

i. *Mode 1*: charging is the term for charging through a standard household outlet with only a simple extender wire in between and no safety protections. Even the house is protected by a fuse but its response in very slow that make unsafe charging. In some countries like U.S.A mode 1 charging is prohibited. So mode 1 charging is much very safe. Clean energy ministerial (2014)

ii. *Mode 2:* charging is refers to charging from a domestic outlet with a cable control and protection device called a cable control and protection device installed in the cable. This method of charging is more secure than the previous one. Mode 1 charging the charging capacity limited minimum rating of household outlet. So the charging capacity is limited it is safer then mode 1 charging. Yusuf, Sk (2018)

iii. *Mode 3*: charging is refers to an EVSE (electronic vehicle charging station) with proper control and protection. This is the most extensively used charging mode on the planet. Yilmaz M (2013) To alternate charging, this can range from 3.8 kW to 22 kW. This is mostly used in charging mode.

iv. *Mode 4*: DC charging is referred to as charging. The on-board charger for electric vehicles is bypassed, and the charging station sends DC electricity straight to the battery through a DC connector. W. Khan, (2017)As a result, there are no needs for an on-board charger or converter in the car.

Required for smart charging system

* In a now day's electric vehicle demand is rising daily in the Indian car industry and the increase burden on the grid it will result in a voltage and frequency imbalance Prabhakar S, (2016).
* The smart grid concept is based on the development of two-way communication between consumers and utilities. Wu H, Shahidehpour (2016).
* When the demand is at its lowest, low peak hours of the night are the optimal times for home-based electric vehicle charging with connectors. S. Srdic (2019).

Conclusion

I would like to suggest for reduces and control the pollution of India peoples have encourage about in electric vehicle. India should place assets into little scope to manage Instead of making major modifications, the local electric burden issues. Chargers at home should be set up for long battery life and structure changes. Great planning of the people, traffic thickness conduct and the security should be consider prior to charging for a wide range of services foundation a death sentence for the world's second most populous country. A formulation of industry best practices for electric car charging is the most important viewpoint for understanding how the electric vehicle-charging infrastructure operates. Electrical essentialness and transport fields. Simply purposeful improvement of the two systems will provide a constant and dependable electrical power structure, particularly at high levels of captive power or believable mind-set of the electric vehicle in India promote. Improvement of vehicle to grid idea alongside grid to vehicle G2V execution for huge scope would help in managing pollution and grid related issues. Yilmaz M, (2012).

Future scope

In this paper, concluded level 3 charging is best among the other charging methods level 1 and level2. According to the pollution graph, the pollution caused by transportation is the most. L. Henry (2018) We should use electric vehicle to control that pollution so that the environment does not pollute. Infrastructure of electric vehicles can we changed when there is development in the sector of charging stations, research and development in the area of battery management systems. Vagropoulos (2014) In future, the road map to the development of charging stations need to be upgraded with the involvement of various development agencies.

References

Chan, C. C. (2007). The state of the art of electric, hybrid, and fuel cell vehicles. *Proceedings of the IEEE*, 95(4), 704–718.

Chen, T., Zhang, X. P., Wang, J., Li, J., Wu, C., Hu, M., and Bian, H. (2020). A review on electric vehicle charging infrastructure development in the UK. *Journal of Modern Power Systems and Clean Energy*, 8(2), 193–205.

Foley, A. M., Winning, I. J., and Gallachóir, B. Ó. Ó. (2010). State-of-the-art in electric vehicle charging infrastructure. *In 2010 IEEE Vehicle Power and Propulsion Conference*, (pp. 1–6). IEEE.

Hossen, M. M., Rahman, A. S., Kabir, A. S., Hasan, M. F., and Ahmed, S. (2017). Systematic assessment of the availability and utilization potential of biomass in Bangladesh. *Renewable and Sustainable Energy Reviews*, 67, 94–105.

Ilieva, L. M. and Iliev, S. P. (2016). Feasibility assessment of a solar-powered charging station for electric vehicles in the north central region of Bulgaria. *Renewable Energy and Environmental Sustainability*, 1, 12.

Jahan, S. and Habiba, R. (2015). An analysis of smart grid communication infrastructure & cyber security in smart grid. *In 2015 International Conference on Advances in Electrical Engineering (ICAEE)* (pp. 190–193). IEEE.

Karkatsoulis, P., Siskos, P., Paroussos, L., and Capros, P. (2017). Simulating deep CO2 emission reduction in transport in a general equilibrium framework: the GEM-E3T model. *Transportation Research Part D: Transport and Environment*, 55, 343–358.

Lee, H. and Clark, A. (2018). Charging the future: challenges and opportunities for electric vehicle adoption.

Musavi, F., Eberle, W., and Dunford, W. G. (2011). A high-performance single-phase bridgeless interleaved PFC converter for plug-in hybrid electric vehicle battery chargers. *IEEE Transactions on Industry Applications*, 47(4), 1833–1843.

None, N. (2016). Power System Challenge: Synthesis Report for the 7th Clean Energy Ministerial (No. NREL/BR-6A50-66482). National Renewable Energy Lab. (NREL), Golden, CO (United States).

Prabhakar, S. and Febin Daya, J. L. (2016). A comparative study on the performance of interleaved converters for EV battery charging. *In 2016 IEEE 6th International Conference on Power Systems (ICPS)*, (pp. 1–6). IEEE.

Sarker, M. R., Dvorkin, , and Ortega-vazquez, M. A. (2016). Optimal participation of an electric vehicle aggregator in day-ahead energy and reserve markets. *IEEE Transactions on Power Systems*, 31(5), 3506–3515.

Somayaji, Y., Mutthu, N. K., Rajan, H., Ampolu, S., and Manickam, N. (2017). Challenges of electric vehicles from lab to road. *In 2017 IEEE Transportation Electrification Conference (ITEC-India)*, (pp. 1–5). IEEE.

Srdic, S. and Lukic, S. (2019). Toward extreme fast charging: challenges and opportunities in directly connecting to medium-voltage line. *IEEE Electrification Magazine*, 7(1), 22–31.

18 Money-saving practices in the rural area of South Wollo in the case of Tewlederie district ACSI branch

Sandeep Kumar Gupta[1,a], Deepa Rajesh[1], Suraj Prasad[2], Sedigheh Asghari Baighout[3], Gurpreet Kaur[4], Neha Nagar[5] and Iryna Mihus[6]

[1]AMET Business School, AMET University, Chennai, India

[2]Central University of Jharkhand, India

[3]University of Tabriz, Iran

[4]Noida Institute of Engineering & Technology, India

[5]Noida International University, India

[6]Scientific Center of Innovative Research, Estonia

Abstract

This research was done on the Tewlederie district of the Amhara regional state for analyses to be carried out and access to information was chosen because of its potential and access to socio-economic infrastructure. The gap identified from other studies is saving cultural differences in different rural areas of Amhara, which are directly reflected in saving practices. This research has been critical in linking saving activities and issues of income diversification to enhance the community's socio-economic status. The research findings have confirmed that households' clients, knowledge of the available source of funding, technical knowledge of farming, and futuristic view of life, display money-saving and income diversification improvement with these indicators of outsourced farmland is insourced and the in sourced live stokes are outsourced by purchasing their own.

Keywords: Amhara credit and saving institution, elasticity, Ethiopian people revolutionary democratic front, food demand, income generating activities, marginal propensity to save, ministry of finance and economic development complement, peasant associations, QUAIDS model, self-help saving groups.

Introduction

Money-saving practices have been started before the birth of Jesus Christ in the Roman Empire. Their story is very much related to the origin of money that traders used to hold (save) their rare metals with the goldsmiths. When they return after their venture, who would give them their gold back because those precious metals have used a means of trade that time and the traders would forego usage. The community of saving has grown from this adage (Butler, 2006; Holston et. al., 2020). In general, formally and informally savings have developed through the negotiation between surplus and deficit of money holders (O'Donnell, 2012;

[a]skguptabhu@gmail.com

DOI: 10.1201/9781003450917-18

Rutherford, 1999; Shuaibu, 2015). Some countries like the USA have concluded that capital support in different forms is important to escape from poverty and, therefore, established a policy that subsidized asset accumulation and it also designed a policy to help the poor to accumulate assets through individual development account (IDA). There have been also a saving theory that is relevant to IDA (Bernini and Pellegrini, 2011; Sherraden et. al., 2003). Japanese evidence reveals that due to mental accounts of potential savings, the median tendency to spend out of income from planned semi-annual incentives is much smaller than for normal income (Ishikawa and Takahashi, 2010).

Gebrehiwot and Wolday (2006) indicated that the government should have clear policies and strategies that could be reflected in all development and sectoral policies in the Ethiopian context to promote the industry. Households, groups, enterprises and institutions can be considered as economic actors in the context of saving practices. The five-year growth and transformation plan (GTP) is still taking the agriculture sector as the backbone of the country's economy. The government indicated in the plan is that saving mobilization is one of the problems in previous plan implementation. The growth and transformation plan which demands huge capital, and the plan implementation would be financed from domestic sources which have been based on saving but it has been identified as the problem and it is also taken as a future resource to finance the upcoming GTP plan; saving has not been incorporated in agriculture sector policy and strategy. Therefore, from the above empirical studies, the current situation of money saving practices in the world can be divided into two broad categories/ well developed and poor money saving practices.

According to Ikhide (1996), a contribution has been seen in the systemic and institutional limitations of savings mobilization, which primarily argues that the mobilization of rural savings is slow because of the low participation of formal institutions. At three stages, the shortage of savings facilities causes problems: (i) the individual level; (ii) the financial institution level; and (iii) the national economy level. The absence of sufficient institutional savings facilities at the individual level requires the individual to focus on in-kind savings such as cash, animal or manufactured goods savings, or unofficial money markets such as Rotating Credit and Savings Associations (ROSCAs) or money-keepers. Saving and livelihood activities of the targeted house are part of the conceptual framework of the study and they are very much interrelated. Rural households with diversified livelihood income could have a higher chance to get better income whereby cash saving and subsequent diversification is possible. Moreover, the study showed that only those who have been considered to be better off have been practicing saving by investing in livestock and cash crop production. Saving was not reported from Danki but instead borrowing from a relative. The study has indicated that borrowing has been made from the relative till the credit service delivery started by government credit association based at Ankober area of Amahara regional state. Further, the panel data which was collected from 19942000 also confirms that the low saving practice has been the very nature of very poor and subsistence (Hirons et. al., 2018; Pankhurst and Bevan, 2004).

Therefore, from the above empirical evidence, it can be concluded that the poor saving and income source diversification is the basic problem in the study area/Tewlederie district/like many parts of Ethiopia. This is due to poor saving culture and livelihood diversification in different rural areas that are directly reflected and strongly linked in saving practices. Thus, the identification of research interventions are cultural and livelihood differences of the study area from elsewhere studied before. The major objectives of the study were to examine the contribution of savings to livelihood diversification in the Tewlederie district, the status of money-saving culture in the rural area of the target district, and the contribution of household income to money savings and their relationship.

Methodology

This descriptive research is based on primary and secondary data collected from local administration offices and people of Woreda Tewlederie rural areas. We have collected a cross-sectional survey of the target population From January to December 2020 among saving clients of ACSI. On the whole, the district is found to have potential in many aspects that provide better opportunities to invest in different fields. where socio-economic improvement could be changed easily within a short period due to the development potentials of the district which is a comparative advantage for any interested organizations and/or institutions. Primary data were collected from 2015-2020, from the district administration office during February 2020.

The actual sample size for the study should be taken from this target population by employing a systematic random sampling technique. Different methods of data collection are included in the design to gather primary and secondary data from their origins, which will be effective with a minimum investment of effort, time and resources to deliver full knowledge. It minimizes bias and maximizes the precision of the results. The research operations the relationship of study variables and the general situations of saving practice in the district has been described with the saving culture of the Woreda residents in a rural area. Independent, the household's income in the research area of targeted institution /independent and saving practice in the countryside dependent.

Sampling design and procedure

Simple random sampling had been employed during the study in which target sampling units from geographic subdivisions of area office had been selected which consists of 51 branches, cluster sampling technique used to select the sampling unit single branch. The reason that's why cluster sampling was preferred was the sampling units which were geographically very scattered and needs more time, money to take the sample from each branch. The selected branch sampling unit of the area office represents all geographic subdivisions whereby the studied trait and study target could be compatible even though geographically very scattered. Therefore, Tewledrie district of Amhara regional state was selected randomly as the research area because it has been known for the existence/availability of private and government financial institutions that have been played the paramount role to households money-saving practice whereby they have been contributed for the livelihood diversification and economic base expansion in the study area and also it is accessible to infrastructures for the researcher.

Systematic sampling technique was employed to take the required sample size 190 clients from a complete list of all 9243 clients sample frame of the selected sampling unit. All individual members of the source list have one fixed chance to be included in the sample without the discretion of the researcher as shown below. N = population of the sample frame, n = the sample size that is included in the study.

Then, K = N/n is the sampling interval and selecting the number between 1 and k/j is a random number to pick up the unit with which to start the sample at the first. Therefore, the sample membership can be determined as j, $(j+k)^{th}$, $(j+2k)^{th}$, $(j+3k)^{th}$...etc. It is the most practical way of sampling to select every th respondents from a complete list sample frame.

The researcher must determine an optimum sample size from a complete list of all clients, sample frames of the selected Branch to increase reliability and accuracy. Therefore an optimum Sample was determined as follow: the population standard deviation is unknown, the desired confidence level of 95%. The expected margin of error is 5%. Population size is 9243 then if the population size is greater than 10,000 sample size (n) = z.p.q/d^2. When the population size is less than 10,000, sample size (nf) = n/1+(n/N) nf = 196/1+(196/9243) =190 .Where n is desired sample size, d is the degree of accuracy, p is the proportion in the target population estimated, N is the population size. Thus, the actual sample size is 190 (Alpha training mogul 2002, 112).

Household and key informant interview (structured or/and unstructured), observation is a well-known data collection tool in which d study was applied to collect the required information from 190 members of customers to describe the general context of the study traits.

Unstructured interview employed unstructured questionnaire containing several open-ended questionnaires whose wording and the order can be changed at will and it helps to get maximum information (opinions) as much as possible and allowing respondents to formulate their answer the way they want and also indicated that open-ended questionnaire gives greater insight and understanding of the topic researched but it was difficult to classify and quantify. Thus, the interview was administered face-to-face with the help of local guides to collect the required data from the targeted (selected household clients).

Data reliability and validity are some of the criteria to accept or reject the findings' conclusion while making a decision. Thus, the conducted research has tried to assure data reliability requirement by taking the following measures. Sampling procedures were transparently outlined and proper collection techniques were utilized. Data collection methods were tested before the actual survey to assure the capacity of the questionnaire in delivering intended information and some of them were modified before collecting the data. The data collectors were oriented intensively in data collection procedures and technique with a close field checking/supervision that helps to give immediate correction at the field level even if, they were oriented well numbers of interviews/forms to be filled per day were minimal per data collector to improve the data quality (Chen et. al., 2002). Moreover, the representatives of the population/sample are selected with no biasness in which each member of the population has an equal opportunity to be included. Representatives of the sample were the vital source of data collected, then reserve respondents were selected in advance and took appropriate action when selected respondents do not present due to their reason. The researcher with MFI experts built up a friendly atmosphere of trust and confidence between the respondents and data collectors in which they felt comfortable, discussed openly, appointed a convenient time to gather the intended data.

Results and discussion

The relationship between savings and income

The most important livelihood asset in the rural community is farmland. To analyze the relationship between these two variables, 190 sample saving clients are categorized into four classes depending on the size of farm landholding. The data gathered from each class was analyzed and explained by:

Table 18.1: Analysis summary of the data.

Classes based on farmland size	Equation (y = a+bx)	r. value
farmland size greater than 0.5 hectares	Y= -2635+0.333x	0.322
farmland size equal to 0.5 hectare	Y= -2099.9+0.261x	0.826
farmland size less than 0.5 hectare	Y=-2563+0.338x	0.346
farmland size is 0.0 hectare/landless/	Y= -2046+0.323x	0.865

This analytical table 18.1 shows that as expected, the relationship between household financial savings and revenue is positive in all classes. The findings support the conventional Keynesian theory of savings, which notes that the relationship between savings and sales is favorable and linear. The positive sign of the expected income coefficient supports the belief that a portion of their income would be set away by family members. Thus, rural households can save in the rural Tewlederie region. The marginal saving tendency, calculated by the approximate coefficient for the financial saving of household savings clients, is positive. The estimated coefficients for household financial savings to income in classes 1, 2, 3, and 4 are 0.333, 0.261, 0.338, and 0.323 respectively. This means that a one Metical (Mt) increase in income, results in a Mt 0.333, 0.261, 0.338, and 0.323 increase in savings in each respective class for financial savings. Therefore, the household's MPS in each class are 33.3, 26.1, 33.8, and 32.3%. Many other forms of research undertaken in developed countries comply with the optimistic savings-income relationship found in this report. A positive savings-income relationship was found in Bangladesh (Siddiki, 2000; Graham, 2011) has also reported similar findings.

The coefficient of correlation/r is a measure of the strength of the relationship between savings and income in each class and all r, values are between 0 and 1/0 < r < 1 indicates that a positive linear relationship exist. But in class 4 respondents the strongest relationship exists between savings and income and in classes 1 and 3 are weak. The finding is surprising that the strongest relationship between savings and income exists for landless clients and clients whose farmland size is equal to 0.5 hectares when compared to others which invite further study.

Cultural status and money saving practices

It is possible to consider culture as a complex mix of tangible factors that can be seen and touched, assumptions/expected attitudes beliefs, customs, practices and people's actual behavior. Each member of the community shares their culture through being and working together. On the other hand, it is also difficult to change because they are 'captured' by their own cultures.

Different nations people have different cultures that influence the practices and/or functions positively or negatively, the issues that someone wants to achieve which can be the basis of competitive advantage in any decision making and action-taking. So, all make and understand the same decisions in different ways. As an agent of change, it is important to recognize and understanding different views that there are significant cultural differences and to consider these differences when creating and implementing any plan. Therefore, women tend to become less knowledgeable and are also sensitive to cultural and religious viewpoints on the kinds of "acceptable" roles for women, and tendency to participate in more "traditional activities that have a much lower profit margin, such as street sales or charcoal production. Females are often reluctant to step out of the constraints of these limited operations because they typically do not have access to structured credit streams. Access by women to structured credit sources is reduced because they often lack collateral, the main source of which island (Fletschner and Kenney, 2014; Sanyang and Huang, 2008). In research, it is expected to review relevant literature to consider the previous and recent efforts that have been made concerning the study topic, and therefore, this research is also trying to review, Imperial, Derg and EPRDF policies and strategies on saving and diversification practices. Moreover, the transformation plan was even reviewed to show the recent measures taken concerning the study title. In general, though employees' credit and saving associations and bank services were started such as Ethiopian Airline and commercial bank, and the regime was trying to boost the country's economy through various economic growth policies and approaches, the national reserve was not enough to finance the plans as the result overall policy depended on external manipulation (outward-looking) due to aid dependency (Woldesenbet, 2020).

Rural household's money-saving cultural practices improvements of the majority can be described from the database established during the data summary. These substantial improvements have been explained under two major indicators. The raising up of saving client numbers within ten years of branch operations and the participant's number of marginalized/disadvantaged groups in the society of the study area. The rise of saving client numbers within ten years of Tewlederie branch operations.

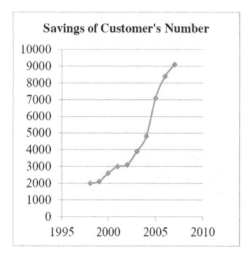

Figure 18.1 Displays the incremental growth of saving customer numbers in the branch within ten years of operations.

Source: Institutions' each year report.

As shown in the Figure 18.1 above the vertical axis represents the number of clients in each saving operations year and the horizontal axis represents saving operation years. Taking the year 1998 as a base year, to compute the range between the base year and the end of the year 2007. Then, 1989-9243 = 7254/364.71% the change of number of clients within ten years. This computing data explains there is a substantial increase in client numbers that reflects the rural household's money-saving culture improved. The marginalized group of womens' money-saving practice in the rural area. Research principle indicates that it is very important to address the issues of a marginalized or disadvantaged group of the community in the study area to link their contribution to socio-economic growths.

According to the study conducted in the Shemene district, women have been involved in productive agricultural activities as a means of livelihood activities diversification. Additionally, 75% of husbands have witnessed that their wives have got a strong desire to make cash savings and 76% of them (Baye, 2017).

Therefore, the theoretical and empirical Studies review is conducted to see whether there has been a gender difference or not on money saving practices among sample households to link the research with others finding. Then study that the researcher tries to see the role of women in money-saving cultural practices in South Wollo of Tewlederie district ACSI branch clients taking as a sample frame. From 190 sample clients, 112/58.95 % and 78 (41.5%) of the respondents are women and men saving participants respectively (Table 18.2). This study tries to incorporate the roles in money saving and livelihood diversification activities of gender in a rural setting by forwarding certain questions to 190 respondents (Table 18.3). Here under-investment needs for their large sums of saved money are summarized that they used it as opportunities to enhance income-generating and living status in several fields of investment.

Table 18.2: Investment, opportunities of women with alternative fields.

Fields of investment	No. of respondents	Percentage (n = 112)
Fattening of livestock	22	19.64
Animal production/husbandry/	18	16.07
Petty trading like mini-cafe hoping, grocery	49	43.75
Crop production	15	13.39
Living house construction	8	7.15

Table 18.3: Investment, opportunities of men with alternative fields.

Fields of investment	No. of respondents	Percentage (n = 78)
Fattening of livestock	13	16.67
Animal production/husbandry/	23	29.49
Petty trading like general chate purchasing and selling, grain, small ruminants etc	11	14.10
Crop production	15	19.23
Living house construction	16	20.51

Source: Own computation from sample frame interview data.

Before 40 years ago our socio-cultural settings in which. The mobility of women, their relationships with members of the opposite sex and their ability to engage in formal education and access to services such as land have been restricted. The outcomes of the summary description are that the improvements of socio-cultural environments in which almost majority of the sample clients are women clients that they can invest their lump sum savings on income-generating activities /IGAs/ to increase their income and to improve living statuses. This clearly has shown that the level of women's effort towards improving their families' income through saving and investment which strengthen their decision-making power but from the selection of their business field and the response of them why they invest in that field indicates that there is mobility restriction of women due the fear of sexual harassment. It is possible to conclude that the non-governmental microfinance of Amhara credited and saving the institution of the Tewlederie branch has been playing crucial roles in encouragement and empowerment of women/the marginalized segment of the society by supporting them to improve socio-economic status. This literature confirms the above explanation.

Rural households money-saving clients' assets and socio-economic status indicators rural farmland

Rural farmland is one of the livelihood assets that could determine livelihood activities/diversification and income of rural households. The total sample frame of the customers (190) can be categorized into four major classes based on farmland assets. 52/27.37% respondents with farm landholding size is greater than

0.5 hectares. In this category, the sampled households farm landholding is even better than other categories of the sample frame. Under this category, the average income and saving per annum are Birr 18992.70. and 3692.40 respectively. In general, from those four classes of respondents with farm landholding size is greater than 0.5 earns greater average income from other classes but greater average savings is greater for respondents haven't farmland assets.

Living house

Housing standards are also used in the majority of developed countries as a surrogate for the socio-economic status measure of a family. Participation in microfinance program has a positive impact on it as well as the level of investment in other sectors like investing in Agricultural inputs to enhance productivity and increasing the yield (Hossain et. al., 2019; Weber and Musshoff, 2012). In the study area of sample frame clients, participation in the ACSI program gives households access to facilitate investments in housing quality. The construction materials used for living house building can determine the quality level of the house. The majority of households in the study are using mud, straw, pole, bark, nails, and stones to construct the wall and also the required materials to construct a roof of the house are grasses and corrugated iron sheets (Wells, 1995). But the quality of the house depends on the quality of building materials they made. The house which is built with the materials such as mud, straw, pole, bark, and grasses are lesser quality than optimum quality building materials like a nail to fasten wood materials and corrugated iron sheets for better roofing of the house which are higher quality and more durable materials. The features of physical housing are a valuable measure of the socio-economic condition of household clients for example number of rooms for corrugated iron sheet roof house is more than grass-roofed houses which is less sophisticated and better in human health conditions better sanitation facilities of the family members which corresponds lower level of health risk. This investment opportunity may help to explain some of the differences in housing quality among society members of the study area. From 190 sample frame respondents, 110(57.89%) customers have made investments for improvements in the quality of their housing in the past five years in the district of Tewlederie. Good housing conditions could contribute further to the betterment of household's health status which will affect the household's productive capacity positively by reducing the prevalence of the airborne disease among households or have an adverse effect due to high chance of disease prevalence whereby it erodes the saving capacity of the households by increasing health expenditure.

Households capacity in terms of economic activities

Debt becomes a liability, and if an intended means of redemption evaporates, the vulnerable sometimes face a crisis. Therefore, to launch new business projects, most individuals must focus on investments. Savings provide potential savings by having access to lump amounts of funds. For business opportunities, such huge amounts of money may be used (Kabeer, 2001).

The aggregate socio-economic activities of the rural households could have implications on the overall performance of clients of the households. For instance, 70 (36.84%) respondents of the sample frame purchases agricultural inputs such as improved seeds, fertilizers, herbicides etc to improve and intensify crop production system to increase the yield, 75(39.47%) respondents in sourcing their outsourcing/renting/farmlands to better-off farmers and 65(34.21%) respondents outsourcing of their contracted live stocks from better-off farmers with the agreements to equal sharing of the offspring and purchased and started their livestock production. These are factors that have positive implications on livelihood diversification, income improvement, asset accumulation and improvements of saving capacity of the poor households. Therefore, micro-enterprises development have been the core business of ACSI saving clients as seen in section 3.2.1.2 and 3.4. For this purpose, ACSI not only provided financial services to its customers but also technical, management, and financial support/credit provision to the success of micro-and small enterprises. The improvements in poor households' economic activities/capacity have a positive effect on the household livelihood security and sustainability as a framework of this study. Institutional preference of customers to save their money.

Financial service has been provided by informal and formal financial institutions for a long time (Tsehay and Mengistu, 2002). Even if, Commercial Bank of Ethiopia has been delivering its service since the Imperial regime. Its role to improve the saving practices of rural society is very weak and insignificant. As Addisu's 2011 study indicated that the informal system could play a saving role where the formal instrument is not better accessed the researcher attempt to link the summarized data with reviewed literature of the study in which the data had been collected from 190 sample frame ACSI Tewlederie branch of rural money-saving clients. From those respondents 153/8, 0.53% of clients prefer to save their money in a formal institution with the justification of its trustfulness/their money safeguarded/and quality service provision/at any time can save and withdraw their savings but 37/19.47% respondents prefer to save their money in an informal institution even if they become the clients of ACSI Tewlederie branch to get loans from the institutions.

Clients formal saving practice and knowledge of money saving practices

Reviewed literature (economic, social/physiological and behavioral theories of money-saving) which are included in the literature. The decision of an individual to save or not is guided by these theories. Knowledge of customers on theories and practices of saving (Ellis, F., 2017) is vital but there are several aspects of consumer awareness, which are the underlying values and behavioral elements that determine how or when customers should save the size and complexity of demand and expectations that push a customer to choose one product or service provider over others. Behavioral economics provides much deeper insight that people should impose resources through systems of mental accounts which have been related to their mental reward and punishment that make it difficult to spend rather than save (Schreiner, 2001). Based on 190 samples of saving clients of the ACSI Tewlederie branch, the contributions of each institution in

Table 18.4: Contributions of each institution for awareness and knowledge enhancement on money-saving theories and practices

Who gives orientations and training?	Number of respondents	Percent of contributions
Agricultural office via development agents /DA/	11	5.79
Amhara credit and saving institution /ACSI	149	78.42
Both Agricultural office and Amhara credit and saving institution /ACSI	15	7.89
Governmental micro and small enterprise offices	11	5.79
Respondents did not obtain orientation and training	4	2.13

Source: Own computation from saving clients interview data.

awareness and knowledge enhancement on theories and practices of money-saving are summarized below.

The findings obtained from awareness and knowledge level testing 0f 190 samples of saving customers interview in the ACSI Tewlederie branch displays 167 respondents from the total samples can respond correctly to the testing interviews (Table 18.4). But 23 respondents didn't respond well to the testing interviews. These data have shown that 87.89% of rural money-saving clients know saving theories, practices and their income source diversification performances like service features, such as deposit, charges, withdrawal terms and conditions, transparency of financial service providers and investment opportunities such as fattening, crop production, petty trading, livestock production, petty shopping's among other to increase livelihood assets with the help of money savings. This is consistent with the mainstream of finance literature (Cole et. al., 2011). Orientation and training programs which have a major influence on uneducated and financially illiterate households to take up and use savings products that establish a strong partnership between financial institutions and customers but 12.11% of rural money-saving clients didn't have clear knowledge and awareness on their income sources diversification and saving practice performances to improve socio-economic status in the community. This group of respondents didn't have strong commitments to continue as clients in the institution and to invest in different income generating activities /IGA to improve their level of income.

Conclusions

This research has given a strong indication to policymaker and local administrations that ACSI has played a strong role in the alternative development of the microfinance sector in which there has been the maximum potential for scaling up. ACSI, is a known successful microfinance institution in the Amhara region involving saving practices. Tewlederie branch office provides services at the grass-roots level to its clients by establishing saving and loan groups in each peasant association. The grouping is depending on the voluntary basis of clients in which the institution using it as the collateral/group guarantee for individual members of the group. The constraints should be addressed by client households, district implementers and financial institutions to motivate rural households for better Scaling up money savings and income source diversification practice the removal of constraints from diversification and expansion of opportunities is desirable for overall objectives of policy because they have been considered that they could give the capability to individual or households to improve their livelihood security and to raise their living standard.

References

Baye, T. G. (2017). Poverty, peasantry and agriculture in Ethiopia. *Annals of Agrarian Science*, 15(3), 420–430.

Bernini, C. and Pellegrini, G. (2011). How are growth and productivity in private firms affected by public subsidy? evidence from a regional policy. *Regional Science and Urban Economics*, 41(3), 253–265.

Butler, S. S. (2006). Low-income, rural elders' perceptions of financial security and health care costs. *Journal of Poverty*, 10(1), 93–115.

Chen, L., Gillenson, M. L., and Sherrell, D. L. (2002). Enticing online consumers: an extended technology acceptance perspective. *Information and Management*, 39(8), 705–719.

Cole, S., Sampson, T., and Zia, B. (2011). Prices or knowledge? what drives demand for financial services in emerging markets? *Journal of Finance*, 66(6), 1933–1967.

Ellis, F. (1998). Household strategies and rural livelihood diversification. *Journal of Development Studies*, 35(1), 1–38.

Ellis, F. (2017). Relative agricultural prices and the urban bias model: a comparative analysis of Tanzania and Fiji. *Development and the Rural-Urban Divide*, 11, 1–22.

Fletschner, D. and Kenney, L. (2014) Rural Women's Access to Financial Services: Credit, Savings, and Insurance, in Gender in Agriculture. Springer, Netherlands, 187–208.

Gebrehiwot, A. and Wolday, A. (2006). Micro and small enterprises (MSE) development in Ethiopia: strategy, regulatory changes and remaining constraints. *Micro and Small Enterprises (MSE) Development in Ethiopia: Strategy, Regulatory Changes and Remaining Constraints*, 10(2), 1–32.

Graham, C. (2011). Does more money make you happier? why so much debate? *Applied Research in Quality of Life*, 6(3), 219–239.

Hirons, M., Robinson, E., McDermott, C., Morel, A., Asare, R., Boyd, E., Gonfa, T., Gole, T. W., Malhi, Y., Mason, J., and Norris, K. (2018). Understanding poverty in cash-crop agroforestry systems: evidence from Ghana and Ethiopia. *Ecological Economics*, 154, 31–41.

Holston, D., Greene, M., and Stroope, J. (2020). Perceptions of the local food environment and experiences with food access among low-income rural louisiana residents. *Current Developments in Nutrition*, 4(Supplement_2), 199–203.

Hossain, M., Malek, M. A., Hossain, M. A., Reza, M. H., and Ahmed, M. S. (2019). Agricultural microcredit for tenant farmers: evidence from a field experiment in Bangladesh. *American Journal of Agricultural Economics*, 101(3), 692–709.

Ikhide, S. I. (1996). Commercial bank offices and the mobilization of private savings in selected sub-saharan African countries. *Journal of Development Studies*, 33(1), 117–132.

Ishikawa, M. and Takahashi, H. (2010). Overconfident managers and external financing choice. *Review of Behavioral Finance*, 2(1), 37–58.

Kabeer, N. (2001). Conflicts over credit: re-evaluating the empowerment potential of loans to women in rural Bangladesh. *World Development*, 29(1), 63–84.

O'Donnell, M. P. (2012). Erosion of our moral compass, social trust, and the fiscal strength of the United States: income inequality, tax policy, and well-being. *American Journal of Health Promotion*, 26(4), 415–420.

Pankhurst, A. and Bevan, P. (2004). Hunger, poverty and "famine" in Ethiopia: some evidence from twenty rural sites in Amhara, Tigray, Oromiya and SNNP Regions. WED Working Paper.

Rutherford, S. (1999). The Poor and Their Money An essay about financial services for poor people, Institute for development policy and management, University of Manchester. https://books.google.co.in/books/about/The_Poor_and_Their_Money.html?id=BHqcIQAACAAJ&redir_esc=y

Sanyang, S. E. and Huang, W. (2008). Micro-financing: enhancing the role of women's group for poverty alleviation in rural Gambia. *World Journal of Agricultural Sciences*, 4(6), 665–673.

Schreiner, M. (2001). Match Rates and Savings: Evidence from Individual Development Accounts. Germany: Center for Social Development, University Library of Munich, (pp. 1–65).

Sherraden, M., Schreiner, M., and Beverly, S. (2003). Income, institutions, and saving performance in individual development accounts. *Economic Development Quarterly*, 17(1), 95–112.

Shuaibu, M. (2015). An empirical analysis of real deposits in Nigeria. *Journal of Economic and Social Studies*, 5(2), 105–124.

Siddiki, J. U. (2000). Demand for money in Bangladesh: a cointegration analysis. *Applied Economics*, 32(15), 1977–1984.

Tsehay, and Mengistu, B. (2002). The Impact of Microfinance Services Among the Poor Women in Ethiopia Occasional Paper. Addis Ababa: AEMFI Publication, (vol. 6, pp. 1–88).

Weber, R., and Musshoff, O. (2012). Is agricultural microcredit really more risky? evidence from Tanzania. *Agricultural Finance Review*, 72(3), 416–435.

Wells, J. (1995). Population, settlements and the environment: the provision of organic materials for shelter. a literature review. *Habitat International*, 19(1), 1–205.

Woldesenbet, W. G. (2020). The tragedies of a state-dominated political economy: shared vices among the imperial, derg, and EPRDF regimes of Ethiopia. *Development Studies Research*, 7(1), 72–82.

19 Industrial automation in food manufacturing

Anand Pratap Tiwari[a], MD Asif Karim Bhuiyan, Sourav Yadav, Anirudh Sharma and Ranjit Kumar Bindal

Electrical Engineering Department, Chandigarh University, Punjab, India

Abstract

The following research paper is cohesively based upon the lights-out process of the industrial automation which is being practiced out there in the industries especially food production; ranging from the initial phase of raw material selection to the final phase of packaging and to a extent somewhere between the manufacturing process. We need less or null human presence or intervention during the entire production process to enhance the overall production quality and retain the overall hygienic standards of the food production for the consumers and sellers out there in the market. The paper is data driven and explains the various technical specifications and compares the applications and regulations of the same as well in various industries i.e., health , education , finance but specially food manufacturing.

Keywords: Automation, embedded, GDP, industry, IOT, lights-out process, microcontroller

Introduction

The pre-existing industrial processes which are currently out there in the market are no doubt efficient enough to procure the demand of the consumers at global level in masses but somewhere lag in the commitment to the hygienic standards and health related aspects in between. May be during the production phase or packaging phase somewhere the hygiene standards get degraded to some extent; and to retain the same, lights-out methodology may be applied on the assembly line during the entire manufacturing process or just at some point of the process in between which means to eradicate or reduce the human intervention during the entire production process.

Literature review

Industries in India

The Indian industries are flourishing and expected to have a dominant presence in the market over the next few decades, particularly in sectors such as processed food manufacturing, textile manufacturing, and medicine-related manufacturing processes. This growth is likely to have a cascading effect on the primary, secondary, and tertiary sectors, ultimately influencing the country's economic structure over the coming decade. As a result, the country's employability rate

[a]anandptiwari186@gmail.com

DOI: 10.1201/9781003450917-19

is expected to increase, bolstering its overall GDP. According to the Index of Industrial Production (IIP), our nation's industrial output rose by 3.4% in April 2019. The country's GDP is expected to be led by agriculture at 16%, followed by industry at 30%, and services at 54%. The government is placing a high priority on promoting industrial growth through infrastructure development, capital access, and research and skill development initiatives, as exemplified by the "Make in India" campaign.

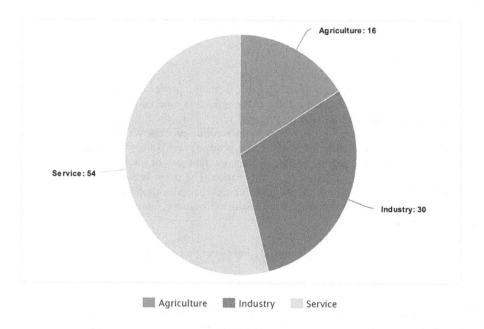

Gaps from the literature

When discussing the inefficiency or underlying causes of lower production rates or reduced benefits, it may be attributed to inefficient industrial systems and insufficient knowledge in the relevant field. Additionally, highlighting the weaknesses in the industrial sector, it is evident that the increasing rate of outsourcing technical equipment is the primary factor that adversely impacts the growth of domestic industries.

Base design implementation

The primary objective is to merge the concepts of embedded systems and high-end technically and financially efficient systems to develop intelligent industrial systems. This will not only improve the overall efficiency of the existing industrial infrastructure but also enhance technical accessibility in the market and expand its reach to all industrial sectors in the country. The conventional industrial practices in our country typically rely on analog switches and error-prone monitoring technologies.

Regulations and design constraints

The project's regulations and design constraints may be influenced by the location limitations and the availability of resources in the vicinity for implementing the project. The design constraints and regulations could comprise the industry's size for deploying the system and the operational environment in which the system must function.

Economic, enviromental, health, manufacturability and safety constraints in design

As an environment-impacting sector, industries produce higher levels of carbon emissions and water pollution compared to other sectors in the country, leading to various health-related issues. However, if we adhere to the necessary government-certified industry guidelines for manufacturability and safety constraints, these issues can be mitigated.

Professional and ethical constraints

From a professional standpoint, the practical implementation of the project in the field necessitates expertise in the electrical and related domains. Ethically speaking, any initiative taken with good intentions and a humane approach that can be utilized for achieving tasks efficiently is deemed as ethically positive.

Social and political constraints

From a social perspective, the industrial sector is a vital component of the economy as it contributes significantly to the country's GDP, making it a socially acceptable practice. Politically speaking, the industry serves as a central ideology and foundation on which political parties develop their policies and discuss the advantages and progress of the industrial sector for their political interests. Therefore, industrial gains could potentially lead to political gains as well.

Technical details

The entire system is controlled using a microcontroller Arduino Uno (Atmega 328p), which is utilized to automate the process by reading sensor data and using it for controlling the environment. The sensor values are used for human intervention in the system.

Multiple design alternates /design solutions

Although there are numerous designs and analog systems available in the market for industrial purposes, the system with the highest output efficiency must be deemed the ideal system design for practical implementation at the ground level.

Pre-crisis estimations

The system design comprises a microcontroller (in this case, Arduino Uno), an ultrasonic sensor for sensing purposes, and an external power supply of 12v connected to an adapter to balance the power requirements on the Uno Board.

Conclusion

The primary aim of the project is to incorporate modern technical efficiency and systems into the current industrial trend. Another objective is to improve the current performance and efficiency of the industrial sector in the country. The main goal is to enhance the overall GDP of the Indian industrial market.

References

Banerjee, M. and Bindal, R. K. (2022). Sensor based solar and wind energy potential and future forecasting in India, in AIP Conference Proceedings, 2555(1), Oct. 2022. [Online]. Available:https://doi.org/10.1063/5.0109187

https://www.researchgate.net/publication/328726221_Industrial_Automation

https://www.researchgate.net/publication/301728845_Industrial_Automation_A_Cost_Effective_Approach_In_Developing_Countries

https://www.springerprofessional.de/en/advances-in-power-systems-and-energy-management

Leong, W. W., Ying, N. S., and Munusamy, N. (2023). Economics: Model Essays, 2nd ed., Scola Books, 2023.

20 Charging of electric vehicle using solar energy- a review

Hemant Kaushik[a], Lalit Kumar Yadav, Shuchi Dave and Priyanshi Mittal

Department of Electrical Engineering, Poornima College of Engineering, Jaipur, India

Abstract

As the demand of electric vehicles (EVs) are increasing day by day because of the environmental aspects as the non-electric vehicles uses the petroleum products to drive the vehicles which are harmful for the nature. The basic requirement of the EVs is the availability of charging stations so that the vehicle can be driven for a long span. There are conventional EV charging stations that use grid power which increases the stress on the grid. So to manage this problem renewable energy based sources may be used to meet the requirements of EV charging. By using the solar energy the emission of SOx and NOx to the environment can be reduced by certain amount. In this paper, a detailed literature review of the various types of EVs and solar based different charging topologies that can be implemented are classified and discussed.

Keywords: Charging methods and charging scheduling, electric vehicles, solar based charging stations

Introduction

In today's era we can see that we are moving towards the electric vehicle (EV) which is driven with the help of battery. EV is like boom for the future, EVs have the bright future for the next generation. While some businesses offer for free usage of internal combustion-based drives have adverse effect on our surrounding. Also the fuel required is based on the fossils which decaying at a very fast rate. So, an alternate is required to meet the day to day requirement of the fuel. As the sun light is freely provided by the environment, so, the solar powered drive could be alternative to fulfil the requirements. Reliability of the EVs is higher in comparison to the conventional drives also they are cheaper as the running cost is comparatively less. As, they provide the smooth drive that's why they are gaining attention day by day. Implementation of EVs basically requires the huge charging mechanism with well supported infrastructure. In this direction a lot theories have been introduced by different people regarding charging mechanism and algorithms. In this paper, a review on the different charging mechanism and MPPT algorithms to extract the solar energy are included.

Different type of electric vehicle

There are four different types of EVs that we are using:

[a]Hemant.kaushik@poornima.org

DOI: 10.1201/9781003450917-20

(a) Battery electric vehicle (BEV)

These type of EVs use batteries as energy storage device and use no other secondary source for propulsion e.g., combustion engine, hydrogen fuel call etc.as shown in Figure 20.1. The electrical energy provided by the battery is then given to the motor to run.

(b) Hybrid electric vehicle (HEV)

This type of EV is the mixture of internal combustion engine and an electric propulsion system. In these EVs the batteries get charged by the generation of power by the internal combustion engine. These vehicles do not require any kind of power source to charge the batteries as shown in Figure 20.2.

(c) Fuel cell electric vehicle (FCEV)

These vehicle works by hydrogen. In these EVs energy is stored in the form of hydrogen gas in a tank and then this energy is converted into electricity to run the vehicle as shown in Figure 20.3.

(d) Plug-in electric vehicle (PLEV)

These types of vehicles use batteries and gasoline tank to run, it uses gas and electricity both as fuel shown in Figure 20.4.

Electric vehicle mechanism

An EV runs using the batteries; the differentials of vehicle are fitted with the BLDC motors as shown in Figure 20.5. These BLDC motors get instruction by the controller.

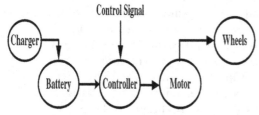

Figure 20.1 Layout of battery operated electric vehicles [Rus et. al., 2019]

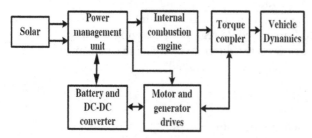

Figure 20.2 Layout of hybrid electric vehicle [Aktaş, 2012]

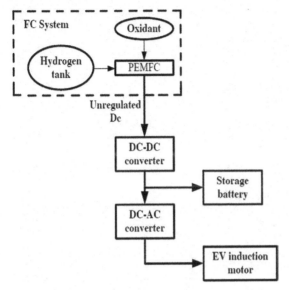

Figure 20.3 Layout of fuel cell EV

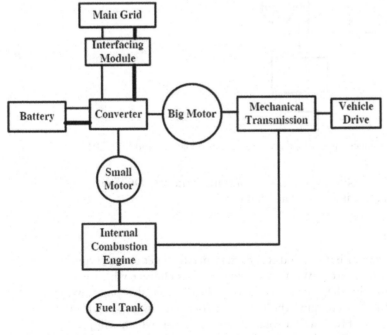

Figure 20.4 Layout of plug-in electric vehicle

Controller also connected with throttle, DC-DC converter. This DC-DC converter converts the 48V to 12V. Controller gets the instructions by the throttle and then it forward to the motor according to the amount of throttle rotate, it contains a metal strip so amount of metal strip passes from it that equals amount

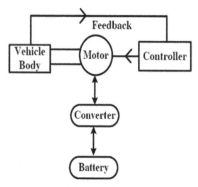

Figure 20.5 Layout of EVs mechanism (Akash, 2020)

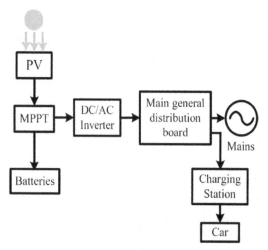

Figure 20.6 Layout of mechanism of solar charging station (Ilieva and Iliev, 2016)

of energy will give by controller to motor. The output of the DC-DC converter connected to the head lights, indicator, back light etc.

Charging mechanism

As EV run using the energy of battery and this battery needs to get charged time to time it can be done using direct AC supply provide to its charger which will convert it into DC supply by the help of rectifier application. In this paper we have discussed the method of charging the EV using solar energy as shown in Figure 20.6. For this we need to install a solar charging station which will produce energy in the form of DC and then this will provide to the battery of EV to charge it.

The amount of energy can be vary so toovercome this problem we need to connect a controller so that it will provide the required amount of energy to the battery. The below given figure is about the mechanism on which solar charging station going to work.

Literature review

The in-depth analysis of the mathematical model for the solar PV array is discussed (Villalva et. al., 2009). By implementing the proposed method, three important points (I, V) can be obtained for a single diode model of solar PV module without any approximation and with minimum details provided by the data sheet by the manufacturer.

A detail comparative analysis of optimum power point tracking methods has been reported (de Brito et. al., 2013). In this article the comparison is done on the basis of the energy obtained from solar PV, tracking with respect to power available, transient response and the implementation of the sensors. PI based incremental conductance method along with the perturb and observe (P&O) method also explained in detail for practical situations.

Maximum-power-point-tracking (MPPT) techniques are implemented on solar PV to obtain the optimum output from solar PV module. Output of the module depends upon the environmental conditions like the surrounding temperature solar insolation etc. Various algorithms are available to obtain the optimum point on the characteristics of solar-PV but in incremental-conductance (INC) method is extensively implemented because of the better tracking capability and better flexibility with the change in the surrounding conditions (Liu et. al., 2008; Kaushik et. al., 2020; Kuo et. al., 2001).

As EVs has a limitation of timing for charging, so the minimization of the charging time is the real fight. In this direction, "allotment of the suitable charging station" to each single vehicle and the charging scheduling are optimized (Das et. al., 2021). In this article, an intelligent charging scheduling algorithm (ICSA) is suggested for the optimization of allotment and charges.

The scheduling of the EV charging is an important aspect, so that each and every partner (supply, electricity provider and the consumer) can be benefited. With this aim, a scheduling plan that optimize the interest of the consumer maintaining the interest of the service provider is discussed in detail (Gupta et. al., 2020).

An offline and online scheduling methods to charge the EV from a solitary station is discussed (Koufakis et. al., 2020). The suggested algorithm is presented to optimize the expenses of charging.

As the requirement of the time, intelligent charging method are critically required to charge the electric-vehicles. In this direction, an adaptive charging network (ACN) algorithm based smart EV charging system is presented (Lee et. al., 2021) to control and manage the charging of EV.

The issue of the global warming motivates to fully accept the inexhaustible energy system based transportation but it's a demanding task. So, a comprehensive analysis of design and implementation of solar-PV based facility to charge a level-2 EV is presented in detail (Shariff et. al., 2020). Different divisions like charging setup, methods considered and control methods for EV charging are also discussed in details.

Personal vehicles and taxis are two common applications of the EV and charging mechanism effects grids remarkably. A novel integrated scheduling method is discussed in detail (Shen et. al., 2021) to manage the load profile for personal

EVs and taxis. The suggested topology consider personal EVs and taxis as a unit to regulate the charging load in total, rather to consider them separately.

A 3-sensor based current controlled protection method to charge on-board EV using the solar PV modules is discussed in depth (Singh and Kumar, 2020). Here the simulation of the proposed scheme is presented using MATLAB-Simulink environment. A hill-climbing method is used to obtain the optimum point and boost converter is implemented to manage the charging current.

A charging unit that includes solar PV with battery support unit along with the utility grid and diesel based generation unit to dispense the interminable supply to EV is presented (Verma and Singh, 2018). All the sources are managed to optimize the running cost of the charging station. In his paper, a voltage source converter is implemented to manage the regulation of voltage and frequency, control the power circulation, reactive power management and harmonic mitigation. The proposed charging method is able to provide DC as well as AC ports to charge the EVs.

Biya and Sindhu (2019) discussed the design of the charging unit including the solar PV and battery unit. A option of utility grid is also attached to give extra support to the system under consideration The proposed method can be implemented to manage the power output for EVs at a units or parking areas.

The growing trend of the EVs has led to improve the present EV charging systems. The system can be considered more efficient if it is supported by the renewable energy options. So, solar PV integrated with the battery and the utility grid based charging unit in MATLAB simulation is discussed (Bose et. al., 2020). The suggested scheme shows that the switching in-between different options of energy supply can be easily done to meet the requirement of charging of the EV.

An adaptive-control based approach has been discussed (Kalla et. al., 2020) for the control of the grid connected solar PV and battery enabled system. The suggested method is found suitable to manage the undeclared-nonlinearity in three phase four wire systems. The suggested control algorithm is implemented to manage the voltage profile at common coupling point and also enhance the power quality for different loading situation by mitigating the harmonics and to compensate the reactive power as well as the neutral current compenzation.

Conclusion

Electric vehicle (EVs) are the one of the valuable vehicle as it provide noise free and create no pollution to the environment. The charging method for EVs are depend upon the configuration of EV. In this article, the detail review of the various papers on the MPPT and different algorithms to charge the battery from solar PV have been incorporated.

References

Aktaş, M. (2012). A Novel Method for Inverter Faults Detection and Diagnosis in PMSM Drives of HEVs based on Discrete Wavelet Transform. *Advances in Electrical and Computer Engineering*, 12(4), 33–38. doi:10.4316/AECE.2012.04005

Biya, T. S. and Sindhu, M. R. (2019). Design and power management of solar powered electric vehicle charging station with energy storage system. *In the Proceedings of 3rd International Conference on Electronics, Communication and Aerospace Technology (ICECA)*, 2019, (pp. 815–820). doi: 10.1109/ICECA. 2019. 8821896.

Bose, B., Tayal, V. K., and Moulik, B. (2020). Multi-loop multi-objective control of solar hybrid EV charging infrastructure for workplace. *In the Proceedings of 2nd International Conference on Advances in Computing, Communication Control and Networking (ICACCCN)*, 2020, (pp. 491–496). doi: 10.1109/ICACCCN51052.2020.9362768.

Das, S., Acharjee, P., and Bhattacharya, A. (2021). Charging scheduling of electric vehicle incorporating grid-to-vehicle and vehicle-to-grid technology considering in smart grid. *IEEE Transactions on Industry Applications*, 57(2), 1688–1702. doi: 10.1109/TIA.2020.3041808.

de Brito, M. A. G., Galotto, L., Sampaio, L. P., e Melo, G. D. A., and Canesin, C. A. (2013). Evaluation of the main MPPT techniques for photovoltaic applications. *IEEE Transactions on Industrial Electronics*, 60(3), 1156–1167. doi: 10.1109/TIE.2012.2198036.

Gupta, V., Konda, S. R., Kumar, R., and Panigrahi, B. K. (2020). Electric vehicle driver response evaluation in multiaggregator charging management with EV routing. *IEEE Transactions on Industry Applications*, 56(6), 6914–6924. doi: 10.1109/TIA.2020.3017563.

Ilieva, L. and Iliev, S. (2016). Feasibility assessment of a solar-powered charging station for electric vehicles in the north central region of Bulgaria. *Renewable Energy and Environmental Sustainability*, 1, 5. 10.1051/rees/2016014.

Kalla, U. K., Kaushik, H., Singh, B. and Kumar, S. (2020). Adaptive control of voltage source converter based scheme for power quality improved grid-interactive solar PV–battery system. *IEEE Transactions on Industry Applications*, 56(1), 787–799. doi: 10.1109/TIA.2019.2947397.

Kaushik, H., Kalla, U. K., and Singh, B. (2020). Modified neural network algorithm based control scheme of grid connected solar PV systems. *International Transactions on Electrical Energy Systems*, July 2020, 31(10), e12547. doi: 10.1002/2050-7038.12547

Koufakis, A.-M., Rigas, E. S., Bassiliades, N., and Ramchurn, S. D. (2020). Offline and online electric vehicle charging scheduling with V2V energy transfer. *IEEE Transactions on Intelligent Transportation Systems*, 21(5), 2128–2138. doi: 10.1109/TITS.2019.2914087.

Kuo, Y.-C., Liang, T.-J., and Chen, J.-F. (2001). Novel maximum-power-point-tracking controller for photovoltaic energy conversion system. *Transaction on Industrial Electronics*, 48(3), 594–601.

Lee, Z. J., Lee G., Lee T., Jin C., Lee R., Low Z., Chang D., Ortega C. and Low S. H. (2021). Adaptive Charging Networks: A Framework for Smart Electric Vehicle Charging, *in IEEE Transactions on Smart Grid*, 12(5), 4339–4350. Sept. 2021, doi: 10.1109/TSG.2021.3074437.

Liu, F., Duan, S., Liu, F., Liu, B., and Kang, Y. (2008). A variable step size INC MPPT method for PV systems. *IEEE Transaction on Industrial Electronics*, 55(7), 2622–2628.

Rus, C., Marcuş, R., Pellegrini, L., Leba, M., Rebrisoreanu, M., and Constandoiu, A. (2019). Electric cars as environmental monitoring IoT Network. *IOP Conference Series: Materials Science and Engineering*, 572, 012091. 10.1088/1757-899X/572/1/012091.

Shariff, S. M., Alam, M. S., Ahmad, F., Rafat, Y., Asghar, M. S. J., and Khan, S. (2020). System design and realization of a solar-powered electric vehicle charging station. *IEEE Systems Journal*, 14(2), 2748–2758. doi: 10.1109/JSYST.2019.2931880.

Shen, J., Wang, L., and Zhang, J. (2021). Integrated scheduling strategy for private electric vehicles and electric taxis. *IEEE Transactions on Industrial Informatics*, 17(3), 1637–1647. doi: 10.1109/TII.2020.2993239.

Singh, H. K. and Kumar, N. (2020). Solar PV array powered on board electric vehicle charging with charging current protection scheme. *In the Proceedings of IEEE International Conference on Power Electronics, Drives and Energy Systems (PEDES)*, 2020, (pp. 1–5). doi: 10.1109/PEDES49360.2020.9379820.

Verma, A. and Singh, B. (2018). A solar PV, BES, grid and DG set based hybrid charging station for uninterruptible charging at minimized charging cost. *In the Proceedings of IEEE Industry Applications Society Annual Meeting (IAS)*, 2018, (pp. 1–8). doi: 10.1109/IAS.2018.8544719.

Villalva, M. G., Gazoli, J. R.., and Filho, E. R. (2009). Modeling and circuit-based simulation of photovoltaic arrays. *In the Proc. of Brazilian Power Electronics Conference*, 2009, (pp. 1244–1254), doi: 10.1109/COBEP.2009.5347680.

21 AC to DC conversion for wind energy on MATLAB simulation

Nayan Roy[a] and Himani Goyal Sharma

Department Of Electrical Engineering, Chandigarh University, Punjab, India

Abstract

Wind power is a very important renewable energy source for electrical power generation. We get wind power from nature, we just need to control this power and convert it to electrical power. Wind turbines and generator are used to get electrical power from the mechanical power comes from the wind power. This is not the end, this electrical power which is produced directly from the machine , permanent magnet synchronous machine (PMSM), or generator, is AC power. This is called rectification. A rectifier circuit converts AC power to DC. In wind generation, it prefers three phase transmission. So , we do the rectification by three phase rectifier. After getting the DC power, we must store it or transfer it for utilization. But presenting of some repulse on the DC, we can get a less efficient power against the wind power grabbing cost. For this solution, here a capacitor filter is used for filtering the pulsating DC. In this paper a MATLAB Simulink model is designed which shows the conversion model of AC to DC power and also how to filter the pulsating DC power for wind generation.

Keywords: Conversion, filtering, generation, rectification, renewable energy, wind energy

Introduction

Wind energy is one of the most important parts of the renewable energy. Wind energy is pollution free, costless, and frequently available renewable energy source. The wind turbines are collecting the wind's power around the environment and converting it to electricity. Wind power generation plays an important role for producing the world's power in a clean, sustainable manner. Wind power is also a domestic source of energy. It harnesses a limitless local resource. If we talk about the sustainable source of energy, wind turbine operation does not directly produce any CO_2 or greenhouse gases. Wind power is helping many countries to reduce their emission targets and fight against climate change. According to the Global Wind Energy Outlook projects, there will be a reduction rate of 2.5 billion tons per year of carbon by 2030. Wind power is faster growing than other energy sources. But there are many challenges to capture pure energy from the generation. The wind turbine produces AC power directly. In order of our necessity, the AC power should be converted into DC power, which is called rectification. There is also a challenge to get the pure form of DC power. This is pulsating DC power including power loss.

In that study we are discussing how to design a wind generation system, how to convert the AC power to DC and how to do filtering the pulsating DC. Here a model is designed in MATLAB Simulink based on AC to DC converter model.

[a]anayanroy17001282@gmail.com

DOI: 10.1201/9781003450917-21

In this model, a permanent magnet synchronous machine (PMSM) produces AC power from the wind turbine with the help of wind power. Then the rectification occurs for converting the AC power to DC (pulsating) and filtering is done at the end.

Rectification

AC to DC converter is presented as rectifiers. Rectifiers are used as independent building management system or correct stand terminal with providing input signal and DC loads. It depends upon the input property of AC system. The speed of response is more than enough to control electromechanical transients which occurring in power supplies and motor drives.

Now we talk about the passive rectifier, which is also known as converters with general commutation or AC to DC line-commutated converters. The supply of the rectifiers would be single phase or three phase supply. To make the circuit simple, the requirements of active and passive components are minimum numbers. Here the commutated power switches are thyristors. The line commutation means the types of commutation. As an example, the transfer of current from one conducting elements to other, the function depends on the main voltage. A thyristor is turned on when a current pulse goes through the gate side.

Three-phase rectifiers

If we need purer direct voltage, then we prefer three phase diode rectifier circuit. Because the three-phase rectifier circuit produces purer DC voltage than single phase rectifier circuit. It also wastes less power [3].

The circuit diagram shows a load connection with three phase rectifier. Where L1, L2, and L3 (three-phase line) are connected to the anodes side of thyristors Vs1, Vs2 and Vs3 via a transformer. At the cathodes side, a load M (motor) is connected. When the L1 gives the positive value, the Vs1 is on forward bias and it makes maximum conduction. On another hand if L1 ison negative alternation, no maximum conduction occurs through Vs1. The other thyristors also operate as the above working process.

The modeled wind generation

The wind generation model is designed on MATLAB 2013a. In figure 21.1 is a model of a wind generation. From SIMULINK library, a blank model has

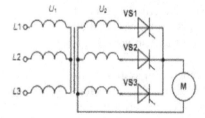

Figure 21.1 Three phase rectifier circuit

been taken. Then a wind turbine is taken on the bland model. The wind turbine is built on hill area or costal area. In figure 21.2 there are three input pots on turbine (generator speed, wind speed and pitch angle). Then we take two constant blocks. One block is for the pitch angle value. The value of the pitch angle is 3 and another block for the wind speed, which value is 100 m/s. Then a PMSM is connected with the output port of the wind turbine. A bus selector is connected with the machine. To get the rotor speed the configuration is fixed for the bus selector as rotor speed parameter. To measure the rotor speed we connect a gain block with the bus selector. The gain block is in radiant form. To convert radiant into sec, we put the value of gain on 30/pi. Then a display and a scope are connected with the gain block. Here the rotor speed is visible on the display.

The PMSM produces AC power, to measure the AC voltage and current, a three-phase measurement block is connected with the PMSM. A scope is also connected to show the VI curve. Now for simulating the model here a load is needed. A three-phase series RLC load has been taken from the MATLAB library. Configure the load with resistance only. Here the value of the resister R=10 ohm. The ground is also connected with the load. Now the primary circuit is done for simulation. When the model is simulated, we get the AC power.

In Figure 21.3, we get the AC voltage and the current curve. The first curve is for voltage and the second is current curve. The machine is running on 238.2 rmp showed on the scope 2 on Figure 21.4. Till now the model is producing AC power from the wind turbine, but we need DC power when we will do rectification on that circuit.

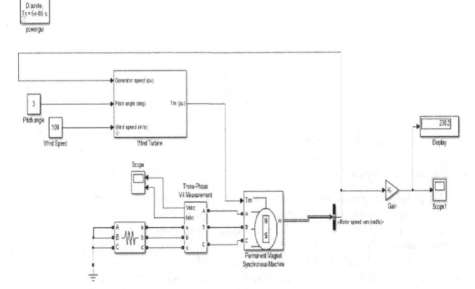

Figure 21.2 AC power generation model

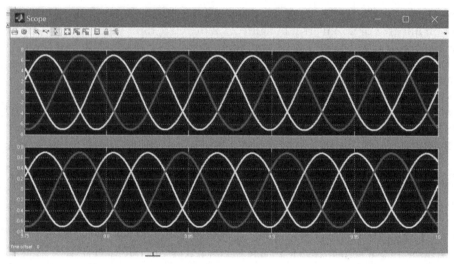

Figure 21.3 V-I curve of AC power

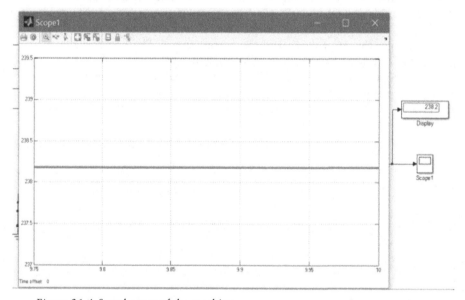

Figure 21.4 Speed curve of the machine

The modeled converter system

We produced AC power on above simulation. Now the AC should be converted into DC power for energy storage. So, to convert the AC power we need a rectifier. To making three phase rectifier we take six diodes from power electronics clock on Simulink library. In Figure 21.5, there is a three-phase rectifier. Where a, b, and c are three phases from the VI measurement block.

Here a resistance R (10 Ω) is also connected with the converter. A voltage measurement block connected with a display and a scope to measure the

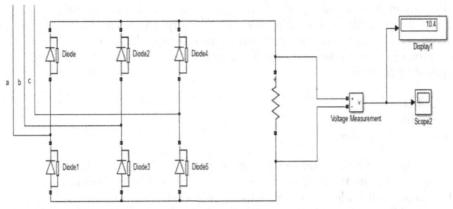

Figure 21.5 Three phase converter model

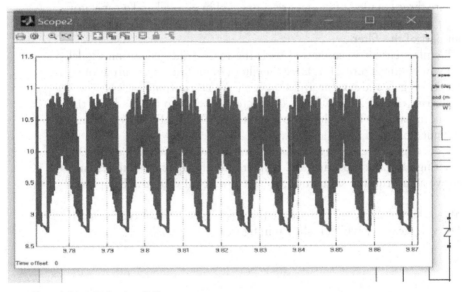

Figure 21.6 Pulsating DC power curve

converted DC power. Now when we simulate the model again, we get the DC power output.

So, we generate AC power from wind turbine and we also converted the AC to DC. But the DC power we converted is a pulsating DC. It has so repulses present on it. Figure 21.6 is showing the pulsating DC power, which is mostly not useable appropriately. To reduce the repulse from the DC, we need to filter the converted DC power.

The modeled filtering of DC power

In previous, a rectifier circuit is used to produce pure DC supply in electronic circuit. But the output of a rectifier is not pure DC, it has pulsations. For this reason, a filter circuit is designed. A filter circuit removes the AC components from the output of the rectifier and produces the pure DC output across the load. The filter circuit is connected in parallel between the rectifier and the load.

Basically, a filter circuit is a combination with capacitors and inductors. There are three types of filter circuit.

I.   Capacitor filter
II. Choke input filter
III. Capacitor input or pi filter

Here we used a capacitor-based filter to filtering the pulsating DC power from the rectifier output. The pulsating output from the rectifier is applied across the capacitor. The capacitor has a low resistance which bypasses the AC components and it prevents the DC component which reaches on the load. As a result on the out there is a pure DC power across the load. In figure 21.7 shows a capacitor filter. There are three coupled of diodes are connected in parallel and a capacitor and a resister are also connected in parallel. The output a, b & c are connected in the input part of this filter.

Finally this filter circuit is connected with the main wild generation model which is showing in the figure 21.8. Here the filter circuit filters the output of so that the distortion gets a low value and we get a pure DC output.

In the converter circuit, a capacitor is connected in between the load and the rectifier circuit in parallel. The value here we put C = 10000 mH. Now our final model is designed on MATLAB Simulink.

So, we produce AC power from a wind turbine, this AC is converted to the DC power (pulsating) and the pulsating DC is filtered with capacitor filter circuit. Now when we again run the simulation, we get low repulses in the DC power curve.

In the above Figure 21.9, we can see that the repulses are reduced then the previous DC voltage curve which is shown in figure.

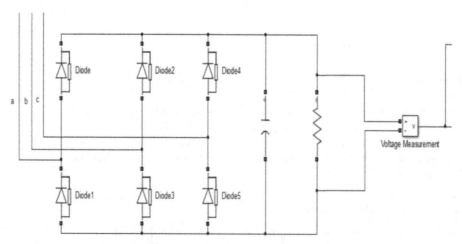

Figure 21.7 Capacitor filter circuit

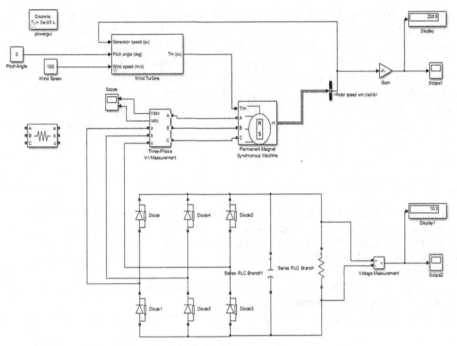

Figure 21.8 Simulation model of wind energy

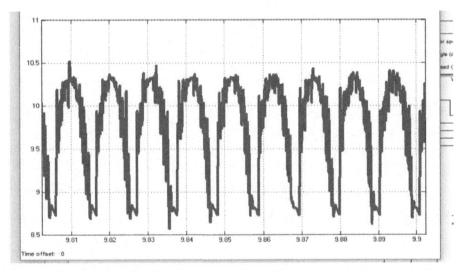

Figure 21.9 Filtered DC power

Conclusion

Conversion system of AC to DC or DC to AC is essential in any generation of plants. In this study we find a general model of wind power generation on MATLAB Simulation. This model presents a three-phase rectifier circuit for the conversion system. Pulsating DC is not very efficient for utilization. We also filter the pulsating DC by using capacitor filter. This model fully presents the basic knowledge of wind generation and the conversion model of wind power. We are living in that time where we use more power than we produce energy. As the limitation of our conventional energy sources, renewable sources are more important. It is also important to generate efficient power from the renewable sources like wind or solar source. As we focus on the renewable energy and the generation of electrical power with more advanced idea and technology, then we reduce economic cost for producing electricity and also help against the global warming.

References

Bajpai, S. and Kidwai, N. R. (2017). Renewable energy education in India.

IEEE 24th Conference, (2009). Annual Applied Power Electronics. (pp. 15–19).

Samantaray, B. and Patnaik, K. (2010). A study of wind energy potential in India.

Yang, K. and Li, L. (2009). Full bridge-full wave mode three-level AC/AC converter with high frequency link.

Yin, M., Lin, G., Zhou, M., and Zhou, C. (2010). Modelling of the wind turbine with a permanent magnet synchro-nous generator for integration.

22 Measuring green IT maturity in businesses

Harshita Virwani[a], Manish Choubisa, Manish Dubey and Nikita Jain

Poornima College of Engineering, Jaipur, Rajasthan, India

Abstract

We couldn't really get back to the conventional approaches to getting things done in an organization, in this manner information technology (IT) associations need to focus on green distributed computing. This paper investigates how any association might have an effect through hypothetical examination (ethnography, subjective exploration with member perception as an unbiased observer). By carrying out green distributed computing administrations instead of standard distributed computing administrations. Total cost to the environment (TCE) can all the more altogether lessen the arrival of greenhouse substances and e-squander created when contrasted with total cost to environment (TCO) while taking a gander at the quantity of fossil fuel by-products delivered by each freely. There might be a gamble to the climate in the event that some IT organizations centre more around TCO than TCE. The discoveries, which think about six aspects — hierarchical, innovative, financial, ecological, social, and showcasing — can coordinate the reception of reasonable tasks and green IT across the organization. The system makes it conceivable to construct, analyse, and assess the association's current green IT rehearses. Likewise, it fills in as a kind of perspective while searching for and monitoring green IT best practices that might be applied to raise the maintainability levels of hierarchical tasks.

Keywords: Corporate social responsibility, eco-ICT, green computing, green design

Introduction

Regulations and environmental movements are putting increasing pressure on businesses of all sizes and in all sectors to lessen their environmental impact. Initiatives in corporate sustainability and corporate social responsibility (CSR) can help to address these issues (Larsen et. al., 2022).

Information technology (IT) is viewed from a strategic perspective as being crucial for every firm that wishes to execute and manage its activities properly and function effectively in its markets. Investments in the purchase and upkeep of technological equipment are continuing to increase at an enhanced rate as a result of this growing technological dependence. However, while IT has given organizations a competitive edge, or at the very least prevented a competitive disadvantage, it has also contributed significantly to many of the environmental issues society is currently facing. Examples include its high energy consumption (which also increases greenhouse gas emissions). In this framework, green IT can be seen as an alternative to traditional innovations that are premised on overall

[a]harshita.virwani@poornima.org

DOI: 10.1201/9781003450917-22

organizational strategies that target the production and use of cutting-edge software and hardware while taking sustainable strategies into account to minimize carbon emissions and ecological consequences (Ajmal et. al., 2018). Examples of studies concerning environmentally friendly IT include the IT lifecycle analysis (LCA), which measures the environmental impact of a product's development, production, utilization, and final disposal—either by end customers or by organizations.

Green IT must take on a central role in order to produce successful outcomes in terms of sustainability. This is because it is the first step to attaining organizational sustainability and making a contribution to sustainable development. Due to its ability to increase global energy productivity while maintaining a dynamic economy, green IT is regarded as the most recent sign of sustainable business practices.

An underlying demand from companies that provide IT solutions from a wider viewpoint, incorporating sustainability goals in other economic sectors in contrast to the institution's IT sector, is predicted for the future. Analyzing the evolution of green IT may be helpful for determining a company's tendency and aptitude for producing sustainable advancements that encompass not just their technological area but, more broadly, the entire organization from an institutional point of view. In reality, there isn't a lot of IT research addressing how to gauge how sustainably established an IT company is. The majority of maturity models are either extremely general or only emphasis on one issue, and thus do not explicitly incorporate the components required to gauge the sustainable maturity level. They accomplished this by analyzing several sustainability maturity models, particularly two IS models, that were discovered in previous studies. This article makes an effort to fill such gaps by describing the current situation of green IT practices in organizations, considering how they pertain to several analytical categories (dimensions and areas), and providing a specific maturity paradigm for green IT in companies.

The study aims to provide a response to the following research question: How can green IT practises be systematised to assist businesses in addressing and diagnosing the growth of IT sustainability in particular organizational areas? Furthermore, we recognize that this framework can serve as a road map for businesses to follow in their search for and monitoring of green IT practices that will increase the sustainability of their operations.

Introduction to green IT

Information and communication technology (ICT) green IT, also known as green computing or eco-ICT, strives to lessen the environmental impact of ICT systems while also enhancing sustainability through their use. Edge computing and other energy-efficient designs, a reduction in the use of rare or dangerous elements like mercury, lead, and chromium in production, and encouraging the recycling of outdated systems are all ways to increase IT sustainability (Nuber et al. 2018).

Green IT is an excellent illustration of how to advance green technology while simultaneously enhancing the state of industrial technology as a whole. This is so that business operations can be significantly improved by increased digitalization and the use of IT systems with technologies like machine learning and data analytics, which in turn increases sustainability (Savaneviˇciene et. al., 2021). Due to the good effects, it will have on the sustainability of the firm as a whole, this can be called eco-friendly despite the ongoing demand by enterprises to increase their computing capacity.

Green IT system best practices

We can significantly lower energy use by modifying the way we use computers. Because most personal computers are purposely left on when not in use, individuals waste energy by having them run even when not using them (De Haes et. al., 2016). The heat that computers generate and the additional cooling they require raise the total quantity of power utilized by the company as well as the cost. The cumulative savings for hundreds of PCs in an organization are significant, even though the energy cost savings per PC might not seem like much. To reduce PC energy use, a variety of steps can be implemented.

Computers can be programmed to shut down automatically to a power-saving mode when not in use without affecting performance. According to green IT estimates from the US Environmental Protection Agency (EPA), giving computers a sleep mode lowers their energy use (Singh and Sahu, 2020).

Many individuals leave their laptops on constantly because they mistakenly think that turning them on and off reduces their lifespan. The temperature and total running time of the electronic equipment affect its lifespan (Patón et. al., 2021). By turning it off, you can lengthen the equipment's lifespan by lowering both of these elements. Recent hard drives are designed to perform reliably over thousands of on/off cycles, protecting internal circuitry from power harm caused by switching on and off. As a result, users have the urge to turn off their equipment when it's not being used (Santos et. al., 2020).

The most prevalent types of green IT practices in an organizational context can be identified by research, including energy-efficient computing (both software and hardware), power management, green equipment projects, environmentally friendly, disposal and recycling, green labelling and the purchase of environmentally sustainable products, and the creation of green industrial IT policies that incorporate the organization's overall environmental policy (Thabit et. al., 2021). These endeavours refer to green practices that integrate "organizational" strategies and imply that such a plan might have a beneficial effect on the environment, despite the importance of these actions being presumed and true (Ozturk et. al., 2011).

Maturity models for IT and green IT

A maturity model is meant to serve as a foundation for comparing and contrasting various organizations, as well as for evaluating practice development patterns, projects, and results. It is feasible to forecast an organization's future performance in a particular area or collection of disciplines based on its maturity level. The findings of the maturity study help businesses prioritize their plans for advancing to greater degrees of maturity and evolving. As a result, businesses can employ maturity concepts and models to speed up their development of practices, procedures, and operational procedures in a wide range of strategic domains (Larsen et. al., 2022).

Depending on the kind of organization or set of processes in question, many models concentrate their thinking and measurement efforts in various areas. There are just a few distinct maturity models for evaluating Green IT, though, according to the literature. They also drew attention to the numerous sustainability and IT maturity models that lack scientific backing and exhibit inconsistency. The fact that the Capability Maturity Model Integration (CMMI) is the model that is most frequently used to construct maturity models in the field of green information technology was also noted by the researchers. IT maturity models serve as tools that assist managers in managing technical tasks, standardizing and maintaining the caliber of information produced and maintained in systems within businesses. The models serve as a tool for IT governance, which gives them strategic value as well (AminiToosi et. al., 2020).

Figure 22.1 shows the market size for green businesses in Asia (Mckinsey, 2022). It is therefore evident that there is still a demand in the literature for more in-depth green IT maturity models, which can provide data for evaluation, organizational situational diagnosis, comparative metrics, and aid in the description of both short and long-term strategies to improve company sustainability goals, in this case, caused by the institution's IT area.

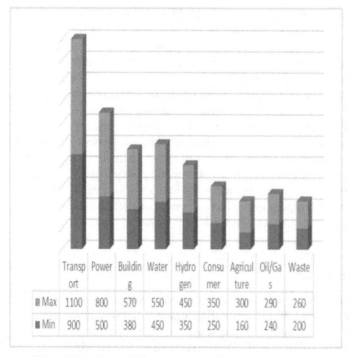

	Transp ort	Power	Buildin g	Water	Hydro gen	Consu mer	Agricul ture	Oil/Ga s	Waste
■ Max	1100	800	570	550	450	350	300	290	260
■ Min	900	500	380	450	350	250	160	240	200

Figure 22.1 Green IT businesses market

Results

In order to make it feasible to assess the maturity of green IT in businesses from a thorough (comprehensive) perspective, this study suggested developing a framework of maturity of green IT. The paradigm makes it possible to diagnose and assess the advancement stage of the organizations' green IT practices as well as the maturity of green IT in specific areas, hence raising the organization's sustainability levels (Collin et. al., 2019).

By category (construct), associated analysis components (block), and final analysis estimate (resulting maturity), as well as taking into account the overall/organizational level of green IT maturity, the framework and potential green IT maturity levels are laid forth.

Figure 22.2 below shows a representation of the framework.

The goal of the green IT maturity framework is to serve as a tool for evaluating the level of organizational maturity with regard to the development of green IT practises from both a top-down and bottom-up perspective.

The most common green IT practise categories used in businesses:

1. Social: An organization's social obligation to people in the future and correspondence objectives that stress everybody's impartial usage of natural assets are the two pieces of the social aspect. This aspect takes a gander at how to accomplish ecological value objectives with an emphasis on guaranteeing that everybody has equivalent admittance to natural assets, an organization's social obligation regarding succeeding ages, the concentricity of

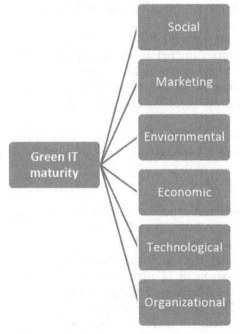

Figure 22.2 Framework for green IT maturity

business objectives with those of the organization's social obligation, the certification of ecological tensions (immediate and roundabout) with partners from both worldwide and nearby populaces, and a foundation's commitment to objectives connected with supportable turn of events.

2. Marketing: The associations outside climate, which comprises of political, monetary, social, and natural variables, is connected to the showcasing aspect. It connects with the activities of associations that should execute evening out procedures, driving the turn of events and reception of green IT, contender activities, interorganizational connections, and market motivators for the reception of green IT, market interest elements that lead to new green IT providers, counselling specialists, and designing branches zeroed in on supportability objectives connected with IT, and the degree of retention of Green IT items/administrations (Dejaco et. al., 2020).

3. Economic: The business prospects that are changed and delivered because of the reception of green IT make up the monetary aspect. To direct green IT projects designated at cost decrease in the medium and long haul, it is expected that there is a sure measure of hazard openness and the accessibility of ventures.

4. Environmental: The ecological aspect is associated with the utilization and the executives of new or further developed advancements with an accentuation on supportability, reception of practices to make existing IT more reasonable, IT items and administrations connecting natural issues and requests in business procedure, affinity for green utilization.

5. Organizational: The reception of green IT practices and advancements (green IS/IT) with an accentuation on proficiency and ecological effect decrease; the securing, reception, use, and the executives of green IT (perfect and proficient IT).

6. Technological: The authoritative aspect connects with the business targets and the essential arrangement of supportability cantered business goals with conventional business targets; IT goals arrangement with key business manageability targets; advancement, reception, use, and dispersion level of Green IT; Green IT the executives and administration; for the evaluation of green IT.

Table 22.1 shows the comparative study of various work in green IT.

Conclusion

The proposed system expects to furnish supervisors and experts with a device equipped for surveying the development of green IT in endeavors in a comprehensive and exhaustive way according to numerous points of view (hierarchical, mechanical, monetary, natural, social, and promoting). The concentrate likewise gives a device to address and analyze the development of IT maintainability specifically hierarchical regions and to coordinate the reception of new green IT rehearses that are presently being presented available. The structure can be seen as a heading for the association's pursuit and observing of green IT rehearses that can be utilized to build the maintainability levels of its tasks.

Table 22.1: Comparative study.

S. No	Author name	Year	Approach used	Findings
1	Larsen et. al.	2022	The paper proposes a deep learning-based approach for monitoring and controlling safety performance in construction sites. Here we used convolutional neural networks (CNNs) and long short-term memory (LSTM) networks, to predict safety performance and provide insights for safety control.	The approach achieved an accuracy of 95.87%, a precision of 95.97%, a recall of 95.98%, and an F1-score of 95.97%, which indicates that their approach is highly effective in predicting safety performance.
2.	Ajmal et. al.	2018	The authors assessed the quality of the included studies using the Cochrane risk of and conducted a meta-analysis of the included studies to estimate the overall effect of MBIs on symptom burden, positive psychological outcomes, and biomarkers in cancer patients.	The paper concludes that MBIs have a positive impact on symptom burden and positive psychological outcomes in cancer patients. The mixed results for biomarkers suggest that more research is needed to determine the impact of MBIs on biological markers in cancer patients. The findings suggest that MBSR and MBCT are the most effective types of MBIs for cancer patients.
3	Nuber et. al.	2020	The authors use of a quality assessment tool allows for a critical evaluation of the included studies, and their synthesis of the findings provides a comprehensive overview of the existing literature. Their proposed future research agenda highlights important areas for future research in this field.	the paper highlights the importance of CSR for firm innovation and identifies the need for a more integrated and theoretically grounded approach to studying this relationship. The authors' proposed future research agenda provides important directions for future research in this field.
4	Radvila et. al.	2021	It uses a qualitative approach to conduct a systematic literature review and content analysis of academic publications related to sustainable development goals (SDGs) and corporate social responsibility (CSR).	It identifies several findings through its systematic literature review and Delphi study, includingthe importance of aligning CSR initiatives with the SDGs for a more effective and meaningful contribution to sustainable development and the need for greater collaboration between stakeholders, including businesses, governments, and civil society, to achieve the SDGs and promote sustainable development.

S. No	Author name	Year	Approach used	Findings
5	De Haes et. al.	2016	It used a quantitative approach to investigate this. The study uses survey data from 54 organizations in Australia to analyse the relationship between IT governance practices and organizational performance.	The findings where its adoption of IT management practices is positively associated with IT operational performance, such as system availability, response time, and problem resolution time.
6	Patón-Romero et. al.	2021	The authors use a qualitative approach to analyse the selected studies and categorize the identified approaches based on their focus, scope, and methodology.	There is a lack of consistency and standardization in the approaches used to assess environmental sustainability in manufacturing, which makes it difficult to compare results across studies.
7	Santos et. al.	2020	The authors use a qualitative approach to categorize the identified studies based on their focus, scope, and methodology. The paper examines the sustainability challenges in the textile and apparel sector and highlights the approaches used to address these challenges.	It found that the textile and apparel sector face significant sustainability challenges and stakeholders are increasingly interested in sustainable products and practices. Sustainable sourcing, eco-design, and closed-loop supply chains are among the most commonly used approaches to address sustainability challenges in the sector.
8	AminiToosi et. al.	2020	The authors identified the benefits and limitations of using bamboo in the built environment. The paper discusses the mechanical properties, fire resistance, and durability of bamboo as well as its potential for carbon sequestration.	This found that bamboo is a promising alternative material for sustainable construction due to its mechanical properties, durability, and potential for carbon sequestration. The paper also highlighted the limitations of bamboo, such as its susceptibility to biological degradation and fire resistance.

This worldview is based on the developmental cut-off points related development organizes that are related with each logical classifications' exhibition/manageability markers, rehearses, techniques, projects, and apparatuses. Furthermore, it contributes by giving a contextualized hypothetical system to green IT rehearses that, when obviously characterized, can decrease deterrents to their reception. This raises examination mindfulness on a layered level and makes green IT application frames more substantial in different settings, which were recently understudied in scholarly world and practice yet fundamental for advancing maintainable changes in the business setting.

The most prevalent types of green IT procedures in an organization context can be identified by research, including energy-efficient computing (both software and hardware), power management, green equipment projects, environmentally friendly, disposal and recycling, green labelling *and the purchase of* environmentally sustainable products, and the creation of green industrial IT policies that incorporate the company's core environmental strategy.

References

Ajmal, M. M., Khan, M., Hussain, M., and Helo, P. (2018). Conceptualizing and incorporating social sustainability in the business world. *International Journal of Sustainable Development & World Ecology*, 25, 327–339.

De Haes, S., Huygh, T., Joshi, A., and Van Grembergen, W. (2016). Adoption and impact of IT governance and management practices: A COBIT 5. perspective. *International Journal of IT/Business Alignment and Governance*, (IJITBAG), 7, 50–72.

Dejaco, M.C., Mazzucchelli, E.S., Pittau, F., Boninu, L., Röck, M., Moretti, N. and Passer, A. (2020). Combining LCA and LCC in the early-design stage: a preliminary study for residential buildings technologies. In IOP Conference Series: Earth and Environmental Science. 588(4), 042004. IOP Publishing.

Larsen, V.G., Tollin, N., Antoniucci, V., Birkved, M., Sattrup, P.A., Holmboe, T. and Marella, G. (2022). Filling the gaps Circular transition of affordable housing in Denmark. *In IOP Conference Series: Earth and Environmental Science*, (Vol. 1078, No. 1, p. 012078). IOP Publishing.

Larsen, V. G., Tollin, N., Sattrup, P. A., Birkved, M., and Holmboe, T. (2022). What are the challenges in evaluating the circular economy for the built environment? A review of the literature on the integration of LCA, LCC and S-LCA in life cycle sustainability assessment, LCSA. *Journal of Building Engineering*, 50, 104203.

Mackenzie, E., Milne, S., van Kerkhoff, L. and Ray, B., 2022, vol 1, Development or dispossession? Exploring the consequences of a major Chinese investment in rural Cambodia. The Journal of Peasant Studies, 1–23.

Nuber, S., Rajsombath, M., Minakaki, G., Winkler, J., Müller, C. P., Ericsson, M., & Selkoe, D. J. (2018). Abrogating native α-synuclein tetramers in mice causes a L-DOPA-responsive motor syndrome closely resembling Parkinson's disease. Neuron, 100(1), 75–90.

Nuber, C., Velte, P., and Hörisch, J. (2020). The curvilinear and time-lagging impact of sustainability performance on financial performance: evidence from Germany. *Corporate Social Responsibility and Environmental Management*, 27, 232–243.

Ozturk, A., Umit, K., Medeni, I. T., Ucuncu, B., Caylan, M., Akba, F., and Medeni, T. D. (2011). Green ICT (Information and communication technologies): a review of academic and practitioner perspectives. *International Journal of eBusiness and eGovernment Studies*, 3, 1–16.

Passer, A., Lützkendorf, T., Habert, G., Kromp-Kolb, H., Monsberger, M., Eder, M., & Truger, B. (2020). Sustainable built environment: transition towards a net zero carbon built environment. The international journal of life cycle assessment, 25, 1160-1167.

Patón-Romero, J. D., Baldassarre, M. T., Rodríguez, M., Runeson, P., Höst, M., and Piattini, M. (2021). Governance and management of green IT: a multi-case study. *Information and Software Technology*, 129, 106414.

Santos, D. A., Quelhas, L. G., Gomes, C. F., Zotes, L. P., França, S. L., Souza, G. V., Araújo, R. A., and Santos, S. S. (2020). Proposal for a maturity model in sustainability in the supply chain. *Sustainability*, 12, 9655.

Savanevičiene, A., Radvila, G., and Šilingiene, V. (2021). Structural changes of organizational maturity during the COVID-19 pandemic: the case of Lithuania. *Sustainability*, 13(24), 13978.

Singh, M. and Sahu, G. P. (2020). Towards adoption of green IS: a literature review using classification methodology. *International Journal of Information Management*, 54, 102147.

Thabit, T., Aissa, S. A. H., and Jasim, Y. (2021). The impact of green ICT adoption in organizations of developing countries. *Al-Riyada for Business Economics Journal*, 7, 9–18.

Toosi, H.A., Lavagna, M., Leonforte, F., Del Pero, C. and Aste, N., 2020. Life cycle sustainability assessment in building energy retrofitting; a review. Sustainable Cities and Society, 60, p.102248.

Printed in the United States
by Baker & Taylor Publisher Services